BITCOIN
FROM BINARY TO GOLD

*YOUR CRYPTOCURRENCY FROM
POOR TO RICH*

HOWARD J. COTE

Table of Contents

Introduction

Welcome to the world of bitcoin and the new digital economy. We will explore the background of this modern currency and explain the reasons behind the highs and lows that have accompanied its rise in popularity.

How will bitcoin affect you in your everyday life? Is this the right time to start using bitcoin or should you wait until it becomes more stable?

We will begin with a brief history and overview of the cryptocurrency and how the association with the Dark Web affected its credibility. Charting the changes in fortune it has undergone since its inception in 2008, we can understand what a remarkable conception this new currency is and just how it is revolutionizing banking and purchases in general.

Bitcoin wallets and how to use them is our next port of call! Safely storing your currency and exploring any pit falls you may encounter. How you can use bitcoin to avoid paying hefty bank charges and exchange rates when you travel. Take control of your own personal finances and explore the potential to bypass banks and other financial institutions that control and monitor your money.

Blockchain is an integral part of the new digital currency usage and is the technology that enables its smooth running. We will examine its properties and find out why such luminaries as Bill Gates and Richard Branson are so excited by the concept of the potential it holds.

Finally, we will come to what is probably the most relevant section, just how does bitcoin and digital currency impact on you and your financial status? What is the future of the digital economy and how soon will it become mainstream? The impact of cryptocurrency on business, government and the society in general!

This book will enable you to understand the basic concept of the bitcoins, how it works, the technology that supports it, and how you can make it work for you. There will be no baffling technological terms just bitcoin for laymen!

Chapter 1:

Bitcoin Overview

a. What is bitcoin

Imagine a way wherein you could anonymously transfer money to anyone instantly, anywhere in the world. Someone is selling something that you want to buy, and you don't want the hassle of transferring funds from a bank account or credit card, or exchanging dollars for pounds or yens – you just want to make an easy and permanent transaction.

If you happen to have a virtual wallet full of Bitcoins you are in luck! Especially if those Bitcoins are more valuable now than they were when you obtained them, giving you more buying power or simply the ability to trade them for a nice sum of cash. Before we get into the nuts and bolts of how to actually make money with this unique form of currency, you will need an understanding of what Bitcoins are and how they are used. You need to know what it is that makes this anonymous currency so valuable and why people are motivated to use it for various transactions in place of many of the various world currencies in circulation.

So what exactly are Bitcoins? Bitcoin is a decentralized, peer-to-peer, virtual currency that was created by a web developer who call himself Satoshi Nakamoto. I say this because Satoshi Nakamoto is widely thought to be only a pseudonym for the mysterious inventor. We'll review the history of Bitcoin in the next section so we're not going to discuss much of it here. But in order to understand the nature of this online currency, it is important to know who invented Bitcoin and why. Bitcoin is the system that Nakamoto invented and Bitcoin is the actual unit of this virtual currency.

The concept of virtual currency has been around almost as long as the internet, providing users a means of exchanging goods and services for currency virtually instantly, without government regulation and with the ability of conducting transactions over great distances and across borders without having to exchange different currencies.

Previous forms of virtual currency had some significant problems, one being that they could be duplicated without value, and there was no real way to verify transactions. For example, one could have an account full of what we'll call "e-bucks" with which one could purchase some goods or services online. However, once the transaction was in process, there was no third party to ensure that the transaction actually took place and actually transferred some intrinsic value from one account to the next. The value of an e-buck literally was based on whatever the holder of the currency or buyer said it was, and what the seller thought it was. There was no way to prevent someone from creating an account full of e-bucks that had no real value. The e-buck was essentially an IOU that could be priceless or worthless depending on the integrity of the issuer of the IOU itself. Let's say I trade an IOU consisting of 10 e-bucks to you for a trinket, and you take those same e-bucks and try to trade them to Joe for one of his trinkets. Joe could question the value of your e-bucks because they came from me and he doesn't know me. Then you could come back to me and demand your trinket back because the e-bucks I gave you are not valuable to Joe. For that matter, let's say that I originally purchased the e-bucks I gave you from an online source for 100 US dollars with my credit card. I could demand that the credit card issuer reverse the original transaction because the e-bucks I bought were no good. The value of any non-commodity backed currency depends upon a consensus of the users on its value; that is, people must agree on what the currency is worth. You can see how complex and ridiculous this can get! Nakamoto developed a system that did not suffer from these problems (though the success of Bitcoin does still require a consensus). Bitcoin transactions are permanent, irreversible, and somewhat anonymous, helping to make them such an attractive form of currency!

Similar to gold, silver, or other types of commodity-backed currency, Bitcoin has a value that is based on supply and demand. And like gold, the supply of Bitcoins is provided by only two sources: people who already have them and people who mine them. That's right, Bitcoins are actually mined, and there is a limited supply to be mined. The effort that goes into mining them and the fact that there is a limited supply are both factors that contribute to the value of Bitcoin. As of the date of this writing, there are around 12 million Bitcoins in existence with a block of 25 new Bitcoins being mined every 10 minutes or so. As the number of Bitcoin increases, that is every

210,000 blocks of Bitcoin, the number created by mining them is reduced by half. At the current rate of mining, Bitcoin should top out at around 21 million in 125 years or so.

b. History of bitcoin

It may come as a surprise to you that Bitcoins were not the first cryptocurrency or virtual currency to be created. There have been others, but not one of them has ever matched the success achieved by Bitcoins. And just like other virtual currencies, Bitcoins also started as an idea.

The very first mention of Bitcoins was in 2008, by a man that goes by the name Satoshi Nakamoto. Note that this is not his real name. In 2009, the very first wallet software, which acted as the system's client platform, was introduced.

The beginnings of Bitcoins weren't phenomenal. Its history is actually muddled by controversy, beginning with the identity of its creator. The price of Bitcoins began where all numbers usually begin – nil. They were usually just traded as collectibles to be had among cryptography enthusiasts.

BITCOIN: FROM BINARY TO GOLD

It took time before its value was appreciated by people. Initially, it was used to trade or purchase trivial things, like pizza and other inexpensive commodities. Its value picked up eventually. And just like any other kind of currency, its price fluctuated up and down through the years.

Some concerns arose, such as the technical glitch in March of 2013. Still, it didn't take long for some mainstream sectors to begin accepting Bitcoins as payment. Businesses, stores, resellers, and service providers accepted this cryptocurrency as a type of legal tender. The very first Bitcoin ATM in the United States opened in Austin, Texas on February 19, 2014. It appears that this virtual currency is gaining popularity in mainstream markets, but it really has a long way to go before it actually gets recognized as true legal tender in many countries around the world.

But one of the biggest controversies to ever hit the system had to do with the online market known as Silk Road. The case revealed one of the weaknesses of the use of cryptocurrencies as a whole. This case proved that such currencies could be used for black-market activities, such as the trade and sale of illegal drugs. At the time, Silk Road was one of the biggest marketplaces on the

Internet that used Bitcoins for purchases and transactions.

Of course, other issues also surrounded Bitcoins, which include theft. Although you basically can't steal a Bitcoin wallet, there are other ways to access the information contained in them. The most vulnerable Bitcoin wallets are the ones that are hosted in third-party sites.

Other threats and issues that came along the way include the Bitcoin-stealing malware, unauthorized mining, and ransomware. All of these have been part of the challenges faced by the cryptocurrency. After all, this is the first time that a kind of cryptocurrency has started to reach into the mainstream. Bitcoins and the Bitcoin community are resilient, and definitely have the makings of success. The difficulties that were mentioned here helped refine the system as it continues to progress forward.

c. Bitcoin uses and their stories

Now that you have your Bitcoins, what should you do with them? There are two main possibilities: use them to purchase things via the Internet or don't touch them and wait for their value to increase.

Make Online Purchases

Quite a few online businesses accept Bitcoin as a form of payment, and in fact, some businesses only accept Bitcoin. Companies that accept Bitcoin include Dell, Wikipedia, Microsoft, Subway, Virgin Galactic, and Overstock. OpenBazaar and BitHappy are two online marketplaces that only accept payments in Bitcoin.

For someone who is familiar with traditional methods of online payment — PayPal, credit cards, etc. — but not with using Bitcoin, the thought of figuring out how to use Bitcoin to process online payments may seem daunting. However, it is actually quite simple. Online companies that accept Bitcoin will have an option under their "Payments" section that will allow you to select Bitcoin as a payment option. You may need to do a quick Google search for how to enable Bitcoin payments for the particular business that you are trying to make a purchase from using Bitcoin. Depending on the business, you will then need to either add funds from your Bitcoin wallet to your account information stored on that business's database, or you can simply link your Bitcoin wallet to that account, much as you would your credit or debit card. Because of the complex process involved in processing a Bitcoin transaction (see "Blockchain: How Does Blockchain Work?"), the transaction may take some time, up to two hours or more, to process.

If you wish to purchase something from a business that does not accept Bitcoins, there is an alternative. Find a website that allows the purchase of gift cards using Bitcoins, such as eGifter.com. Use Bitcoins to purchase a gift card for that business in the amount that you want. You can then use the gift card to make the purchase. While this method is a bit more time-consuming, many people find that it is worth the extra effort because the rapidly increasing exchange rate of Bitcoin is enabling them to buy things that they otherwise would not be able to afford.

On the same note as using Bitcoin to make online purchases, you can send and receive Bitcoin from other people. Say that you want to send your nephew Bitcoin as a birthday present. The procedure may vary slightly according to the wallet that you are using, but you will need to have the Bitcoin funds in your account, as well as his public Bitcoin address. Then you can simply transfer the money to him using his public address, not unlike wiring money to someone using his or her email address.

Investing with Bitcoin

Another thing that you can do with Bitcoin is using it as an investment. What this means is that you don't touch the Bitcoin but rather allow its value to increase, not unlike how you would allow investments in your 401(k) or other investment portfolios to increase in value.

While cryptocurrency is new and there is certainly no guaranteed way of figuring out what the value of currencies such as Bitcoin will be in the long term, Bitcoin's value has increased from one ten-thousandth of a penny in the summer of 2009 to nearly $3000 in the summer of 2017. If this trend continues, Bitcoin will be worth hundreds of thousands of dollars in a few years. Investments that have shown that much gain in a short period of time are usually seen as risky. However, in combination with other investments, Bitcoin can be a meaningful way to build value to your portfolio.

d. What makes bitcoin unique

Of course, to make a fuss in the information space, to get zealous adepts, selflessly devoted to the idea of crypto-currency, and could become a lust of stock-market speculators because it is unique.

Earlier nothing of the kind existed. Bitcoin is a new milestone in the history of mankind.

Having absorbed the well-known achievements of peer-to-peer networks, bitcoin presented the world with a lot of new ones.

1. So, the decentralized network protects the bitcoin from the arbitrariness of the authorities of a particular country or some monster of the UN. It is pointless to adopt a convention on the prohibition of bitcoin, to introduce into parliament a law that places bitcoin outside the law. It is even more foolish to amend the penal code, which provides for punishment for the use of bitcoin (although anything can be expected from the most

democratic countries). The absence of any central server or data center makes it impossible for its storming by Smith agents with the support of unmanned aerial vehicles with guided missiles. The absence of a formal owner in the same way saves the entire network from a sudden fatal blow; And it's exactly like this in May 2013

The popular international payment system Liberty Reserve (LR) was destroyed. The US authorities carried out a raider capture of the company with lightning speed, arresting (of course, others') hands and capturing the server with the database. Users of the payment system in one minute left with nothing, without any hope to ever turn their LR units into something more easily convertible. However, they also lost access to these units. Yesterday there was money - today there is nothing left. In principle, in this way, you can destroy any payment system - But not Bitcoin.

The fact that bitcoin does not belong to anyone and belongs to everyone, puts all users on an equal footing.

In those countries where the prospects of spreading the "no-man's" cyber-currency have already ripened, they try to convince the population through controlled media that bitcoin is bad, contagious, dangerous, and in general, a respectable citizen will never have affairs with bitcoin.

That, of course, only smiles user's bitcoin. In December 2013, bitcoin convincingly proved that, to spite all the intrigues of enemies, people need and remain on the Internet, whoever pinches his teeth. An avalanche of news about the persecution of virtual currency sharply crashed the bitcoin rate by several hundred dollars, but within a couple of days he played $ 200-300, where he stopped with minor fluctuations. Thus, the measures of information impact on cyber-currency through the media apparatus were not very effective. And no other state has any other.

2. No virtual money you can store on your computer; but bitcoin was created just for this storage! In the case of bank accounts, the computer simply give you access via the Internet. Bitcoins are physically stored on a hard disk, flash drive or any other storage

media. And stolen bitcoins cannot dispose of them, unless, of course, you provide him with kindness with a sincere private key of the signature that allows you to perform operations to send bitcoins to other network members.

3. The creation of money (emission) in the real world is unlimited and unlimited. Do you know how much US dollars are issued? Yes, no one knows. Some tens of trillions, to be exact, no economist-expert-analyst will say. The issue of money automatically devalues them. Simply because the market for their use remains the same in size. What do you think, how much will a loaf of bread cost if every person on the planet has a suitcase with two million dollars?.. The number of bitcoins is strictly limited. Will be created 21 million bitcoin. It is this number of "beads" that mankind will have. It is easy to imagine that if every seventh inhabitant of Russia gets only one bitcoin, then all other countries in the world will not get a single coin. Moreover, it can be argued that none of those reading this line will survive until the last generated bitcoin is released. The process is greatly stretched in time and every year it slows down,

that is, every year the number of bitcoin created is less than in the previous year.

4. You know perfectly well that there are fake money. The fight against counterfeiters in any state is fierce and irreconcilable. Punishment for the printing of banknotes, as a rule, is the most severe (in the USSR it was punishable by the death penalty - execution). This is quite natural, since counterfeiters undermine the financial system and the economy of the state (during the Second World War the Germans printed counterfeit Soviet rubles, tossed and fake sterling pounds to the British). In the field of virtual money, all banks are fighting for the protection of their databases, because "draw" an extra million dollars on their bank account - the cherished dream of any novice hacker. Bitcoin also eliminates the problem of counterfeit money easily and elegantly. It is impossible to create a fake bitcoin outside the network, and then "add * to the total turnover. It is impossible to "draw" bitcoins in any database. The total number of bitcoins will be 21 million, and on this issue will end. It is easier for a hacker to come up with a different crypto currency, than something to philosophize in the bitcoin network.

5. Known in advance a finite number of bitcoins will actually be somewhat less. The fact that, as stated in paragraph #2 above, bitcoin is possible (and desirable) to hold under the arm (on the flash drive, for example). However, if you lose the flash drive or it gets corrupted, in short, the information it will be not readable - you will lose your bitcoins. Network cannot provide you coins in exchange for the lost. Each bitcoin is created once, and if it is lost, to replace it new will not be created. Such cases have already occurred, some of them quite tragic. For example, James Howell of England once got on the cheap a few thousand bitcoins (and bitcoin was not always so expensive as it is now - at the dawn of the network they exchanged rather a joke, not even imagining that the cost of the jokes will soar to the heavens). He kept them on the hard drive of the laptop. Time passed, James had forgotten about what was involved in bitcoin, and in General about this topic. Unserviceable hard drive has been thrown away. When the rate of bitcoin topped $1000 for 1 bitcoin, James remembered that he had on a hard drive hidden away six million dollars, and rushed this disc to look for. But the chance to dig up the landfill area at several stadiums and find there safe and sound hard drive, as you know, is zero. So James lost sleep and tear the hair on my ass, hoping to go back and bury the hard drive in the safe. The spice of this story makes the fact that James sort of activity just a geek. A lot of cases where people lose keys of their electronic wallets, losing access to bitcoin. Bitcoins, which lost access, in the same way as physically lost kryptonite cannot be replaced by others. Given the inevitability of such losses, this automatically implies that bitcoin will not be ENOUGH, and after will be created the last "cue", their total number will start to decrease due to the carelessness of the owners. Experienced users are advised to keep bitcoin at least three places.

6.. Each enumeration in the Bitcoin network is recorded in "an open book", so data about the amount of the wallets of sender and recipient is available for everyone at any time. Thus, the usual excuse dishonest taxpayers, they say, "you transferred the money, it's the banks delay", whereas in fact the money is not transferred, - the Bitcoin network is impossible by definition. Sending a transfer, you can immediately email you send the recipient the link to the transaction record in the "open book". Or require such a link, if bitcoins are sent to you.

7. When making transfers between different banks, one can often not be sure that the beneficiary's details are correct, and there is no possibility to check it directly in the sending bank. Even if you make a transaction on the IBAN account number in the European country via the Internet bank, where there is nothing to indicate, except for the beneficiary account number (more accurately, not necessarily), it is safe to say only that the account number corresponds to the IBAN standard, But not that such an account actually exists (unless the sender and recipient are customers of the same bank). In the Bitcoin system, it is possible to determine whether an address exists at any time without even attempting to send funds. If in a traditional banking system funds with incorrectly specified details can "wander" for days, or even longer, then in the Bitcoin network this is impossible by definition.

8. All payment systems that depend on a centralized database are vulnerable to one degree or another. This applies even to the

largest players on the market. The probability of the disappearance of funds from the balance due to a fault in the work or damage to the database is never equal to zero. Money can "get lost" in the transfer. Bitcoin cannot be lost under any circumstances. All that you sent and received, is stored in the memory of many computers in different countries on different continents. Even if an atomic war destroys 99% of computers with this information, the rest will be able to restore the data and distribute it to the rest of the network machines.

9. The uniqueness of the Bitcoin-address allows using it not only in the field of money transfers. Bitcoin-address is a great login! Which can be used when registering for various services? Its advantages over the usual combination of «login + password + confirmation by e-mail», and now also by the code sent by no SMS, are that there is no need to confirm anything. If e-mail can be hacked, "withdraw" letters with passwords and logins, then it's impossible to crack Bitcoin's address. In general, when registering with a Bitcoin-address, you do not need to invent a password and make a confirmation, because it's useless

to register with someone else's Bitcoin-address. Therefore, for example, the fusions (which are discussed further on) are mostly required for use only by the Bitcoin address, and no more body movements. In addition, since each Bitcoin-address is unique, the problem of someone else's login is completely eliminated.

10. Bitcoin is ideally suited for payment transactions for goods / services, which in other electronic payment systems are either impossible, or risk entailing the anger of the law on the participants in the transaction. Yes, we can say that selling drugs for bitcoin is evil, and therefore bitcoin should be is forbidden. But, for example, online casinos from the use bitcoin (and only bitcoin) are convenient to both casino owners and players. The former are insured against the risks associated with payment systems, the latter retain the virgin history of using credit cards. Bitcoin allows you to use any online services to residents of countries in which any of the services are prohibited or access to which is blocked.

Such a small number of bitcoins, which tends to decrease, means that in the course of time the cost of bitcoins in the usual money will only grow. The exchange rate fluctuations are natural as a consequence of the activity of speculative traders, but one can definitely say that there will never be a cheap bitcoin ever. The times when the cue ball was given away for nothing or for a penny, irretrievably gone into the past. Ahead - there are times when you cannot buy bitcoin even for a lot of money. But the owner's bitcoin become very wealthy, wealthy people. Bitcoin is not so much the equivalent of money that you can spend for everyday needs (to drink coffee, refuel the car ...), how much investment, not subject to inflation. Any currency you are familiar with, whether it is the euro or the dollar, is subject to inflation, i.e. Loses part of the value, and depreciates. Bitcoin is not subject to inflation.

Chapter 2:

How Bitcoin Works

a. Birth of Blockchain

We can nitpick about the precise moment in time that Bitcoin entered the mainstream. Generally speaking, however, even though there have been spikes in Bitcoin's value before, most would agree that 2016 – 2017 have marked a significant uptick in mainstream interest and adoption of both Bitcoin and blockchain technology. As a potential investor, it is useful to look at the history of Bitcoin to gain some perspective on how it has evolved. Studying the events and behavior that have shaped the current market can help us shape our vision of what the future may hold and make informed decisions.

Perhaps one of the most anecdotally interesting aspects of Bitcoin is the mysterious identity of its creator. Bitcoin first appeared in 2008 when a paper illustrating the concept was published to an email list for cryptography enthusiasts. The paper, written by a figure known as Satoshi Nakamoto, was called, "Bitcoin: A Peer-to-Peer Electronic Cash System."

A few months later, in January of 2009, Satoshi Nakamoto implemented Bitcoin and released the code as "open-source" (meaning that anyone can look at the code). Nakamoto mined the first block of Bitcoins, sometimes called the "genesis block," which started the bitcoin blockchain.

Although a few people have claimed either to be Nakamoto or to know his/her/their true identity, none of these claims have ever borne out. To this day, the identity of Satoshi Nakamoto remains a mystery. Nobody knows if Satoshi is one person, a group of people, a secret society, or the alias of a nation state actor. What is known is that he/she/they are thought to have mined around a million Bitcoins in the early days of the blockchain. At that time, of course, Bitcoin had a tiny user base and little value as a currency.

One of the most famous early Bitcoin transactions was the exchange of 10,000BTC for a delivery of two pizzas. Today, of course, 10,000BTC would be worth upwards of $20,000,000! At the time,

Bitcoin was mostly seen as a subcultural novelty that a small group of people were beginning to experiment with.

Many people hear about the drastic increase in Bitcoin's value over the past few years and wonder if it is too late for them to make money by investing in Bitcoin. Of course, everybody wishes that they had bought Bitcoin earlier, but speculation still plays a huge role in the Bitcoin community.

Will the price of Bitcoin skyrocket again, will it rise slowly over time, or have we seen the height of its value already? These are the kinds of questions that draw many people to Bitcoin today.

b. How bitcoin get created

i. Bitcoin blocks

All blocks in the main chain are numbered, starting with the number 0, then 1, 2, 3, 4, 5, and so on. The green block is the first block that was created, and it's also known as a genesis block, and it has a block number zero.

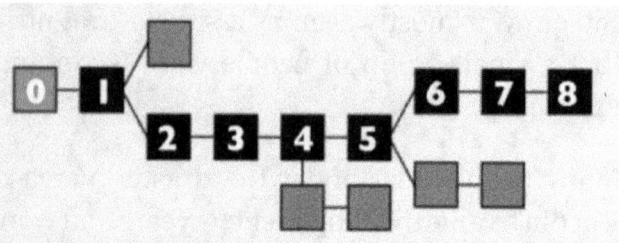

The purple blocks are the ones that are forming short and invalid chains, they are called blockchain forks. Blockchain forks do occur very often, additionally these side forks, also known as orphaned forks.

A Bitcoin block is created every ten minutes on average; however, Ethereum blocks are set up in every 17 seconds on average. The block height is the sum of the blocks in a chain between it and the genesis block minus 1. Blocks on side forks can have the same block height as blocks on the main chain. Particular nodes on the peer-to-peer network are creating these blocks. These nodes are called miners. All the miners are collecting every transaction that people are sending to each other over the network, and only valid transactions are relayed to the other nodes. Each miner takes a number of these collected operations and puts them in a newly formed block. These lists of transactions are numbered tx0, tx1, tx2, ... and so on. Tx stands for transaction, followed by the number. The first transaction (tx0) is also known as the coinbase transaction. This is the transaction where the miner assigns a block reward to his address. This is how Bitcoins are created. For Bitcoin miners, as of now in 2017, the block reward is 12.5 Bitcoins; however, back in the day of the genesis block, the reward was 50 bitcoins for each block creation. For Bitcoin, the block reward is halved after every 210,000 blocks. Once there have been 64 halvings, the block reward will be zero. There will be a maximum number of 21 million Bitcoin in circulation in the year of 2140. Other Bitcoin transactions, such as tx1 or tx2, are the ordinary transaction where the bitcoins are transferred from the owner address to a recipient address. Each transaction requires a small transaction fee. This fee will continue to increase as an incentive for the miners to create new blocks because the block reward will continue to be lowered.

When the miner has constructed the block, he must solve a hash puzzle that is applied on his list of transactions. The miner who first solves the hash puzzle is allowed to broadcast his block on the peer-to-peer network. The block also includes the solution to the puzzle, also called the nonce, in the block header. This is, of course, available to anyone who wants to see it and the details for each block can be found at www.blockchain.info

Other miners on the network will receive this block, and they validate the block before they append it to their chain of blocks. It happens regularly that another valid block is broadcasted on the network because another miner has solved the puzzle nearly at the same time.

When this happens, temporary forks are created. For example, fork A and fork B. Let's assume that 70 % of the miners on the network are working on fork A, and the rest of the miners are working on fork B. In this example, fork A becomes the main chain because it consists of the longest series of blocks from the genesis block. Miners should always work on the longest chain. In this example, blocks on fork B will become orphaned blocks.

The miner who solves the hash puzzle, and his block is on the main chain, will receive the block reward and also all the transactions fees (tx1 and tx2) in this block. The miner who has solved the hash puzzle, and his block is an orphan fork, cannot spend the block reward and transaction fees, because his block is not on the main chain.

ii. bitcoin transaction

All Bitcoin really involves in trading. Since it is not an actual officially accepted currency, people can just trade for Bitcoin and trade their Bitcoin for different things. There are many different ways that you can trade your Bitcoin and different things that you can do. Bitcoin is good for trading everything from cash to homes and even small goods like groceries. If you are going to use Bitcoin for trading purposes, you need to be sure that you are doing it the right way and that you are not losing out on money as a result of the trades that you are making with the Bitcoin that you have in your own possession.

For Services

The majority of services that you can trade for Bitcoin are online-based services. Some people who use Bitcoin will choose to trade it for:

- Website creation

- Content marketing

- Site enhancement

- Computer optimization

Because Bitcoin is a computer-based currency and something that people can trade online, it is something that has a lot of use in the online world. In the past, people who were doing each of these things and trading them for other things would actually have to pay for them using a credit card (which can be risky) or using a payment service, like Paypal (which can be extremely complicated).

The introduction of Bitcoin makes it much easier for people to try and pay for the services that they have online. It is something that everyone is able to benefit from. Even if you don't have a website, you don't have any use for content marketing or you don't want to create a website at any point, you can use Bitcoin to trade for a service that will help make your computer run better than what it ever had before.

By using each of these services, you will enable yourself to have a better computer or technology experience. If you want to be able to make the experience even better, you can choose to trade it for Bitcoin so that you have to take one less step to be able to pay for those services. Bitcoin makes it easy to pay and even communicate with the people who are going to provide those services to you.

For Goods

There are many goods that you can purchase online. The majority of these goods are found at online retailers and many of the biggest retailers in the online world have now started to accept Bitcoin as a form of payment. You can use your Bitcoin wallet at the most popular retailers and that will enable you to have an easier and more secure online shopping experience. To be able to use your Bitcoin at an online retailer:

1. Find what you are looking for and add everything that you want to purchase into your shopping cart or shopping bag on the retailer's site

2. Go to the cart or bag and review the things that you are buying

3. Use the checkout button that is provided

4. Find the place where you can pay with Bitcoin

5. Enter your personal information so that your items can be shipped to you

6. Find the place and enter your Bitcoin wallet ID so that you can pay for the items

7. Confirm the purchase

Your Bitcoin wallet will automatically be updated with the information on the purchase and it will be deducted the amount that you just purchased your items for. You do not have to wait for a statement or anything to clear because it does it all instantly.

Chapter 2: How Bitcoin Works

Traditional Trades

Bitcoin is a great thing to have when you are doing trade investments. These are the type of investments that include things like stocks, bonds, and even mutual funds. You can do the same thing with your Bitcoin and simply purchase them for investment purposes.

If you were going to invest in something else or even if you were going to hang onto the money in the form of traditional cash in a savings account, it would take years to build up a return on the total amount that you put into the account. The percentages of even high-interest savings accounts are much lower than the return that you will get on the Bitcoin that you invested in.

Another great aspect of trading in an investment sense with Bitcoin is that you do not have to put it in a specific account to build up the value of it. All you need to do is allow it to sit in your wallet that the rest of your Bitcoin is in. This will allow it to build up in value as the Bitcoin continues to grow, as a whole, in value. Within a year, there is a chance that you could make back up to 25% (or more) of your investment.

When You Purchase

When you buy your Bitcoin, you need to make sure that you buy it at a good time. On the weekends and in the early morning before the market opens for the day is the best time to buy your Bitcoin. This is something that you will need to do to make sure that you are getting the best price. Watch the prices of Bitcoin for a few days and see when it is the lowest.

If you buy your Bitcoin at the lowest point within five days, you will be able to automatically make a return on that investment so that you can make sure that you are actually making money from the Bitcoin. It is a good idea to only buy it when it is as low as possible. If you think of the people who initially bought Bitcoin, they likely only spent a few dollars for Bitcoin that is now worth millions of dollars today.

Buying at the right time will have a huge impact on your ability to profit with Bitcoin.

When You Sell It

Opposite of when you are buying Bitcoin, you need to watch out for the highest price when you are selling it. You also need to keep in mind the selling fee if you are trying to sell it on a site like Coin Base because you could end up losing money on the sale if you don't do it at the right time. By selling it when it is at its highest, you will be able to make the profits that you need to be able to buy more Bitcoin and turn around to do the same thing.

If you sell your Bitcoin when it is at the highest price possible, the chance that it is going to drop again is good and it is something that you need to remember when you are selling. After you have sold it, monitor the price for a few days. Use the profits that you made from the sale to buy more of the lower-priced Bitcoin. This is the easiest way to make a lot of money from Bitcoin and it is also the fastest. There is no long waiting period in between when you buy and sell the Bitcoin so you can make profits fast.

Swapping Bitcoin

While you already know that you can trade Bitcoin for goods and services, you can also trade Bitcoin in a different way on the Internet. If you are using Bitcoin, you can trade it for other Bitcoin, for different types of cryptocurrency or even for different things that you have. For example, if you have something that is very unique or rare, you can actually use it to barter for Bitcoin.

Since Bitcoin is not an official currency and it is not regulated, you can use it for almost anything. That means that you can ask someone for three Bitcoin for a book that you have and they want. There is no guidelines for trading or for selling things with Bitcoin so keep that in mind when you are the one who is trying to get Bitcoin. It can be complicated to figure out what you should pay or what you should accept as payment.

When you are using Bitcoin, always be careful because it is not regulated. While you can get a lot of "money" for things that aren't quite worth that large of an amount, you also have the chance of getting ripped off with Bitcoin.

Always know the value of things that you are buying and selling as well as the value of Bitcoin at that point in time.

iii. bitcoin mining

Now that we have a better idea of how bitcoin works, it is time to move on to how you can get started with bitcoin. There are a few different options, but we are going to talk about the process of mining. This is a good one to use if you want to get bitcoin and you have a little bit of experience when it comes to working in coding. When you work with some of the traditional sources of money, you will find that the government is in charge of printing more money when it is needed, but when it comes to bitcoin money, you will need to discover it when you want to create some more. There are a lot of computers all over the world and these can be used in order to mine the coins. Here we will take a look at some of the steps that you need in order to mine the bitcoin and when you are able to discover these coins, you will have some of your own.

How Does Mining Take Place?

In order for there to be more bitcoin, you will need to mine them. There is the maximum of 21 million of the bitcoin, but not all of them are in use right now. You may need to mine them in order to get more to go into the whole system. The process of mining is pretty simple, but you will need to work on it a little bit and it takes some time, which is why not everyone is going to choose this method.

Now, there are many people all throughout the world who will use this bitcoin network to send some of the bitcoins between each other, for making purchases and more. But without having some kind of record of these transactions, it is hard to keep track of which accounts have made the payments or not. Inside of the bitcoin network, there is a process for collecting the transactions that were able to happen during a set period, usually during one day, and then this information is going to be placed on a list.

For the system to keep track of the various transactions that will occur, which can often be a large list each day, a miner will be able to go through all of these and then confirm the transactions. They are able to take each individual transactions before writing them down onto a ledger.

The Next Steps

So at this point, you may be uncertain about the point of the ledger. Why would you as a miner be so interested in learning this ledger and keeping track of it? The general ledger is going to be a long list of all the blocks that you use in this system. You are able to use this list in order to explore all of your transactions that are made between any of the addresses that you would like.

So on the blockchain, you are going to see all of the transactions that show up each day. As time goes on, this list will start to get fairly long because it will have lots of transactions that will show up over time. Anyone who is inside of the bitcoin network is going to be able to get a copy of this report so that they can decide to be a miner if they would like.

In order for this system to work, the general ledger needs to have some kind of process so that all of the users feel that their information is safe. This is where the miners are going to come into play. They are going to make sure to change the ledger in such a way that the information will stay safe and others won't be able to get the information.

When a new transaction block is created, it is going to be the job of a miner to put it through a process to keep things safe. They will help to take this information and then place it into a formula so that the information will be turned into something that is brand new. This is going to give you the results of something that seems random and is a lot shorter. You are basically going to include letters and numbers that are considered a has in this system. The hash will be stored inside of the block, somewhere near the end of the blockchain that you created.

You are going to notice that these hashes are able to have some cool properties. You can take some of this data inside the bitcoin block and it is pretty easy to make a hash from that information, but since this information will be random, it is almost impossible for others to come out and see what the data means. Each hash that is created will be unique and the nice thing is that they are able to take a big amount of data and give them their own unique hash so that it doesn't match up at all. For, example, if you are able to change up just a single character inside of the hash, you are going to be in a brand new transaction and it will look completely different.

Miners are not just taking the transactions in order to generate the new hashes. They are going to be able to use a little bit of information that is inside this data in order to help out. One important piece that you can use is the last block that is stored right at the end of the blockchain.

The neat thing is that each part or character of the hash is going to be used in order to produce the following one, you are going to get something completely unique each time. This helps to keep the security on these transactions because this method will help to create a wax seal inside of the digital world. The system will be able to look at the block, and the other blocks that are with it, and then if someone tries to tamper with the code, everyone inside of this network would find out quickly.

So if you are trying to be a miner and tried to fake one of these transactions, or you are trying to fake some for another reason inside of the system, you are going to have some issues. Even just changing one of the blocks that is in this chain is going to change up the whole block. If someone tried to see if the block was authentic with the hashing function, they would be able to see that something was changed and it is easy to see if the transaction is one that is false right away.

Since all of your hashes will be used with this same idea, with the character of one determining what the character of the next and all down the line, it is easier to see if someone is messing around with all of them, which helps them to stay safe.

It is the job of the miners to help keep track of all these different transactions inside of the system for bitcoin. They are going to be able to use an approach that is randomized and that can help to keep these transactions as secure as possible. In fact, if the job is done in the proper way, you will be able to see that all of the transactions will stay safe.

How To Get The Coins

Now that we understand a bit about how hashes inside of the system and why all of this security is needed, it is time to work on what exactly a miner would do in order to compete and get some more of the coins. The miners are going to be the ones who are responsible for sealing off a block of the transaction with some randomized hashes so that they can compete with each other, using different types of software that has been designed for mining. And when they are able to create one of these blocks successfully, they are going to earn 25 bitcoins.

This is a great win for everyone. It is going to help those who are inside of the system stay safe ad secure so that people don't mess with the transactions or send out information that isn't true. It also helps the miner out because if they are successful, they are able to make some good money from these. Because of the high amount of money that you can make from these, you will find that many miners will work to create some more of these hashes and it can become very profitable.

The biggest issue is that the bitcoin network had to work to make things harder to create. In the beginning, people were able to just get a computer program and they were able to make money, but since this was so easy for a lot of people to do, so bitcoin had to make it a big harder to do.

There is now a process that is used that known as a proof of work that has helped it to become more difficult to create the hashes so that not all of the money in bitcoin could be mined in just a bit of time. The protocol with this network is going to help you to make sure that you meet everything when creating one of these hashes so that you can make some money. First, you should ensure that the hash looks a certain way in order to have a set amount of zeroes at the beginning. Since there is no way for you to know which way the hash will look until it is all done, you may have to make quite a few of them before you get it to work. And any time that you add in some more data to your system, it is going to change the hash up.

As the miner, you will make sure that you aren't going through and meddling around with the information that is with your transactions, but you will also need to make some changes in order to get your hash to become created and meet up with the rules. They are able to bring in another type of data, which is also going to be random so that you are able to create something that is known as a "nonce". This is going to be used as well as the data for the transaction so that you can make a brand new hash. So if you find that the data isn't fitting into the hash, the bitcoin will let you use the nonce so that you can change your hash around as well.

As you can imagine, the new rules make it so that you have to do quite a few tries to make sure that everything is going to work. But when you are willing to work with a computer program and keep on going with it, you can find that it is possible to make quite a bit of money with this mining option. The system is offering you $25000 USD for each of these hashes you are able to create, which means that if you are able to stick with it, it becomes possible to earn some good money. Make a few of these a year, and you have a good income.

Many people who are good with computer programming and like working with this kind of thing find that it is a great way for them to pick up a challenge and make some good money. If you like this idea, it is one of the best ways to go with to make some bitcoins without having to sell a product or do anything like that. Make sure that you have a good computer program that is able to help you to get started on this whole process of creating your own hashes.

c. How bitcoin network gets maintained?

As we know, bitcoin mining works on a basis of proof-of-work. The framework is designed to depend on the previous block to force a chronological order in the block chain this this will make it impossible to reverse any previous transactions because this would require re-calculations of the proof-of-work of all of the subsequent blocks.

Assumedly if two blocks are discovered at the same time, miners will work on the first block(first in, first out) before switching to the next block. This mining method will provide a global consensus throughout the entire network.

Bitcoin miners therefore are not able to cheat or increase their bitcoin rewards through fraudulent transactions as it would corrupt the nature of Bitcoin network as all Bitcoin nodes would reject any block that contains invalid data.

It is often acknowledged that, once all the Bitcoins have been mined, the miners that maintain the Bitcoin network are compensated with transaction fees. Ideally, this is the main incentive for miners. However, this approach can kill one of the main competitive advantages of the Bitcoin crypto—which is lower transaction fees.

Decentralization is (and will still continue) to be the main fuel behind the Bitcoin network. Without decentralization, many of the Bitcoin characteristics such as facilitating transactions without involving a third party or providing a permission-less platform for innovation can be compromised. Obviously, there are many facades that contribute to the Bitcoin decentralization. One such element is the network of nodes that make up the Bitcoin's infrastructure.

The Bitcoin nodes maintain the network by storing copies of the Blockchain and sharing the block together with the transaction data across the entire network. But despite this property and its significance in the Bitcoin network, the number of nodes has been declining steadily over the years. One can argue that this has an effect of centralizing the network.

During the early phases of Bitcoin development, the only way that you could participate in the network was by running full nodes. Over the years, the system flourished and now there are several wallet options that users can select from. Majority of the wallets are either light-weight clients that are querying full nodes for information or they are hosted by third-party institutions and therefore don't require running a full node.

Because of this, the majority of new users are choosing against running full nodes, while some existing node operators have completely shut theirs down. I know you're now asking, "How many nodes does a bitcoin network really need?"

Well, depending on the perspective that you look at Bitcoin maintenance, you can reach several conclusions:

- One: Because bitcoin is a trustless network, the only node that is important is the node that you run.

- Hundreds: Or enough to make it untenable for any single node to shut down a significant portion of the ecosystem due to geographic and jurisdictional diversity.

- Thousands: Or enough to support high demand from the SPV clients for the connection slots. SPV clients may not necessarily be wallets, but can also be peer-to-peer applications.

On the opposite end of the argument, it is impossible to have too many nodes that can decentralize the network. Having said that, how will the Bitcoin network be maintained if less than 1% of the bitcoin users run a full node?

One argument that often crops during the block size debate is based upon the cost of running a node. There is a theory that higher costs (such as additional computational resource requirements to validate and relay larger blocks) will result in fewer nodes and vice versa.

Obviously, the lower costs should result in more users undertaking actions that make business sense to them. Technically speaking, there are no defined minimum resource requirements for running a full node. Therefore, there is no target for the Bitcoin developers to take into account when exploring the potential of making the protocol change to result in increased resource requirements for running a full node. If a minimum specification is to be implemented, it should be based on the current hardware that is being used to run the full nodes.

For instance, an ARM-based device such as ODROID+ or Raspberry Pi appears to be the current minimum possible specification to run a node. It can keep up with the 1MB block sizes, though it can take up to two weeks to perform the initial blockchain synchronization because of the low-powered CPU.

A well-designed specification should be set with the parameters of performance characteristics that are desired for a node, the resources required to meet those minimum performance targets and the cost of obtaining the hardware that meets the performance parameters.

For instance, a minimum specification can look like this:

- Target hardware cost: $200

- Target worst-case time to validate the block: 10 seconds

- The minimum network I/O: 2 MB/s

- The minimum disk I/O: 2 MB/s

- The minimum CPU: 5,000 MIPS

- Minimum RAM capacity: 1 GB

Obviously, the goal of keeping up with the minimum specifications is to keep the node operation costs low and accessible to an average user. On the other hand, if you maintain the resource requirements of the nodes at the level of your resources, the transaction costs may increase.

Put simply: if the cost of running the network increases to the point of excluding average users from making transactions on the Bitcoin's blockchain, then they are not run the node at a lower cost. You can think of this as the balance between the cost of transaction validation and the cost of the transaction.

There are several costs to run a full node, such as:

- The initial learning curve (time cost)

- The installation, configuration and the initial sync cost (time, CPU and bandwidth)

- Ongoing running costs such as CPU, bandwidth, RAM and hard disk

- Maintenance costs such as time to perform upgrades and troubleshoot the network.

The initial time cost (learning curve) to see the value of bitcoin may take weeks or even months while the initial sync time can take several hours to several weeks depending on your hardware specs. While we have examined the cost of running a Bitcoin node from a variety of perspectives, it's only sensible to theorize that higher costs can result in fewer nodes participating in the network and lower costs will result in higher nodes.

d. Bitcoin Nodes type and their roles

A bitcoin node is a collection of 4 functions and they are;

a. Wallet

b. Miner

c. Blockchain

d. Network

To function fully, the bitcoin network must not only rely on one avenue for transactions but also require other selected nodes for preventing double spending(eg, user double spending the digital currency twice)

In general concept, the more bitcoin nodes is available in the network, the more the node type, the more secure the network is.

Currently, most full bitcoin networks nodes are maintained within the North America continent while some other continents are lacking of such resource. This lack of nodes in several continents is due to the reason that participants are not rewarded with any monetary benefits when a user or organizations set up a full node unlike participants who mine bitcoin will get awarded accordingly based on their contribution

To view the live map of the current available network nodes, users can log into https://bitnodes.earn.com/

Members of the Bitcoin development community understand it's difficult to maintain bitcoin nodes while not incentivizing is not self-sufficient however they believe that major companies within the bitcoin industry will continue to pick up the role of bitcoin nodes as they understand the vast implications behind to maintain a full node.

Now that we have explored the process of maintaining a Bitcoin network, let us now turn to Bitcoin node types and their roles. But first, let's explore the concept of Bitcoin nodes.

As was mentioned, Bitcoin was conceived as a peer-to-peer (P2P) network on top of the web. By peer-to-peer, I mean computers that participate in the system are treated as equal and there are no cases of 'special" nodes. These nodes interconnect together in a complex mesh network without a centralized server and no hierarchy within the network.

The nodes in a P2P network both consume and provide simultaneously with the reciprocity acting as the incentive for participating in the network. P2P networks are inherently decentralized, resilient, and open.

So, what are the different types of nodes that exist in a Bitcoin network and what are their roles?

Even though the nodes in the Bitcoin P2P system are equal, they may assume on different roles depending on the functionality that they are offering. A Bitcoin node can be regarded as a collection of functions that may include:

- The Blockchain database

- The routing nodes

- The mining nodes

- The wallet services.

Let's dive in to explore these nodes.

#1: Blockchain database

Blockchain nodes—also called full nodes— maintain an up-to-date and complete copy of the

Blockchain. These nodes can authoritatively and autonomously verify any transaction without an external reference. Some of the nodes maintain only a subset of the blockchain database and verify transactions using Simplified Payment Verification (SPV). The SPV nodes (also called lightweight nodes) don't have the full copy of the Blockchain database.

#2: Routing Nodes

Routing nodes validate and propagate transactions and blocks while discovering and maintaining connections to other peers. All the nodes in the Bitcoin ecosystem have this function besides other roles such as blockchain database, wallet nodes, and mining nodes.

#3: Mining Nodes

The mining nodes compete to develop new blocks by executing a specialized hardware to solve the POW (proof-of-work) algorithm. Some mining nodes can also be full nodes, maintaining the full copy of the blockchain database, while others may be lightweight nodes participating in pool mining.

#4: Wallet nodes

The user wallets can be part of the full node, as is the case with the desktop bitcoin clients. Increasingly, the majority of the user wallets, especially those that run on resource-constrained devices such as the smartphones, are SPV nodes.

The Extended Bitcoin Network

The main bitcoin network (that I have just described above) runs the Bitcoin P2P protocol, with between 7,000 and 10,000 listening nodes executing various versions of the Bitcoin client (the Bitcoin Core) and a few hundred nodes that execute various implementations of Bitcoin P2P protocol that include Libbitcoin, BitcoinJ and Bitcoin Gold.

A small percentage of the Bitcoin nodes on the P2P network are also mining nodes that compete in the mining process, validating the transactions, and generating new blocks. Various organizations interface with the Bitcoin network by executing full-node clients that are based on the Bitcoin Core client, with the full copies of blockchain database and the network nodes, but without elaborate mining or wallet functions.

Such nodes can act as network edge routers, allowing multiple other services such as exchanges, block explorers, wallets, and merchant payment processing to be implemented on top of the normal Bitcoin nodes. These nodes create an extended bitcoin network includes offers the functionalities of the normal Bitcoin nodes and specialized protocols such as:

- Reference client. It has the wallet, the miner, full blockchain database, and the Network Routing Node running on a P2P network.

- Full blockchain node. It contains a full blockchain database and the Network Routing Node on the P2P network.

- Solo miner. It contains a mining function with a full copy of the blockchain database and the routing nodes on the P2P networks.

- Lightweight (SPV) wallets. They contain a wallet and the network node on the P2P networks.

- Pool protocol servers. These are gateway routers that connect the Bitcoin P2P networks and other protocols.

Chapter 2: How Bitcoin Works

Chapter 3:

Current Bitcoin Market

a. How is bitcoin valued

When Bitcoin was first brought on as a type of cryptocurrency (one of the first, actually), it was not worth much. Many people did not think that it was going to last and some even thought that they would lose money by investing in it. Others, though, believed in the idea of it and they invested in it. Those people were smart because the Bitcoin that they invested so little in just eight years ago is now worth a lot of money. For just a few dollars, they are now millionaires.

Year 2009

Average Bitcoin Price: .0001 American Dollars

If you owned 10 Bitcoin, you would have: .001 worth of Bitcoin, less than one cent

Years on the market: 0

The people who first invested in Bitcoin did not pay much for the amount that they had. In fact, people who invested 10 dollars actually got 100,000 Bitcoin. They did not know at the time but they would eventually be able to see a huge return on that money.

While 10 dollars would have gotten you a lot of Bitcoin during that time, people who really believed in the idea behind Bitcoin actually invested a lot more than just 10 dollars. Some invested hundreds, even thousands of dollars into Bitcoin. They didn't know at the time but they were actually helping to increase the value of Bitcoin and inflate the price of it by doing this.

Year 2010

Average Bitcoin Price: .07 American Dollars

If you owned 10 Bitcoin, you would have: .7 worth of Bitcoin, just under one dollar

Years on the market: 1

Chapter 3: Current Bitcoin Market

This was the first of many huge increases in Bitcoin in the past eight years. It started out at far less than one cent and the price was driven upward by investors and those who wanted to be able to get in on the Bitcoin game. This is the point that many people saw that Bitcoin may not be such a joke after all. While the seven cent price point was not huge and was actually far less than some of the other investment options that were on the market, it was a huge increase.

Good investors were able to realize this and started to invest even more money into Bitcoin. The average purchase of Bitcoin often involved thousands of dollars. The people who were doing this were not big-name investors but were, instead, technology aficionados who wanted to be able to get in on the action.

Still, at this point in time, if you had 10 dollars to spare, you could get yourself 142 Bitcoin. Those who did this during that time were wise.

Year 2011

Average Bitcoin Price: 15 American Dollars

If you owned 10 Bitcoin, you would have: 150 dollars' worth of Bitcoin

Years on the market: 2

This was the second time during the lifespan of Bitcoin that it took another huge jump in price. It was something that was propelled by the investors who were spending thousands of dollars to be able to get into the Bitcoin game and it was something that also allowed them the chance to make sure that they were going to be able to have a lot of money later on.

By this point, most investors realized that Bitcoin was going to stay on the market. They knew that it was something that had the potential to change the way that the stock market worked in a huge way and that it was going to be able to be one of the fastest growing investment opportunities in the world.

Chapter 3: Current Bitcoin Market

Year 2012

Average Bitcoin Price: 7 American Dollars

If you owned 10 Bitcoin, you would have: 70 dollars' worth of Bitcoin

Years on the market: 3

For the very first time in its short history, Bitcoin dropped in price. Compared to the huge increases that it had in the past, this was a relatively large dip but it was something that was to be expected considering that Bitcoin had only been on the market for a short period of time. In fact, the 7 dollar price point was something that many people thought the Bitcoin would peak at.

People began to back out. They sold off their Bitcoin in fear that it would drop again in the coming months.

The smart investors, though, knew that there was always a drop in price before there was another huge increase. They held onto the Bitcoin. They waited it out and the patience that they had for the market that they had chosen to join paid off for them.

Year 2013

Average Bitcoin Price: 100 American Dollars

If you owned 10 Bitcoin, you would have: 1,000 dollars' worth of Bitcoin

Years on the market: 4

Those who held onto their Bitcoin during the drop in price were rejoicing because they knew that they had made the right decision by keeping the Bitcoin that they had. It was important that they kept it because it allowed them to have a much larger amount of Bitcoin.

While these people felt that they were able to be blessed because they had a lot of Bitcoin, those who had purchased it back in 2009 were even better off. If you had purchased 10 dollars worth of Bitcoin in 2009, you would have been a millionaire in 2013. There were actually more than a few people who reached that point. Many of them cashed out at that point – satisfied with the fact that they had made a million and wanting to pull out before they lost money. Still, others stayed. Many figured that it couldn't get much worse than that initial 10 dollar investment.

Year 2014

Average Bitcoin Price: 600 American Dollars

If you owned 10 Bitcoin, you would have: 6,000 dollars' worth of Bitcoin

Years on the market: 5

This was another increase but not a huge one that had been seen in the early days. While it was smaller than those times, it was a big increase and it propelled those had bought during the 2009 Bitcoin selling season to be multimillionaires.

Again, more people cashed in on the Bitcoin that they had. They wanted to get out before it was too late. This was especially true of those who had multi-millions of dollars at stake.

Year 2015

Average Bitcoin Price: 220 American Dollars

If you owned 10 Bitcoin, you would have: 2,200 dollars' worth of Bitcoin

Years on the market: 6

During this time, the illegal market that was running on the Internet, the Silk Road, was seized and closed. It was a time that hurt a lot of people who dealt exclusively in Bitcoin and it caused the price to drop.

The seasoned investors knew that this was bound to happen so they did not let it get in the way of their investments. They wanted to continue cashing in on Bitcoin so they did just that. They left their Bitcoin in their wallets and did not sell just because of the drop. They hoped that it would rise again.

Year 2016

Average Bitcoin Price: 1,146 American Dollars

If you owned 10 Bitcoin, you would have: 11,460 dollars' worth of Bitcoin

Years on the market: 7

For the first time in a long time, Bitcoin more than tripled in value. Those who had bought 10 dollars worth of Bitcoin in 2009, now had a wallet that was worth over one hundred thousand dollars. The thing, though, was that people who had bought all of those shares up in 2009 actually spent thousands of dollars on Bitcoin. The year 2016 was celebratory for them in that they passed the billion dollar mark. While they have it all divided up into different wallets, they are still billionaires.

2017

Average Bitcoin Price: 992 American dollars (as of January)

If you owned 10 Bitcoin, you would have: 9,920 dollars' worth of Bitcoin

Years on the market: 8

While there was a slight drop at the beginning of the year, it is expected that Bitcoin is going to pass the price of gold in the coming months. It is expected to hover around the $1,000 mark for a few months and then shoot up to nearly $2,000.

There is no way to determine the exact amount, but some people expect Bitcoin to reach a value of over $10,000 by the time that we reach 2020.

If you haven't already, now is the time to buy Bitcoin so that you can cash in later.

b. Advantage of bitcoin over regular currencies

A few of the benefits brought by Bitcoin are seen in effective markets. A Bitcoin can be divided into millions of parts (every part is called satoshi); the fiat currency is normally broken down in hundreds). The transactions in this network are free, or in some cases include a tiny transaction fee in order to induce the miners. But we are speaking of approximately a tenth of one percent. If you are to compare this with a two or four percent fee that is generally charged by the credit card companies, you will understand why this concept is so attractive.

If you want to participate in this economy, you do not have to be a technical expert or to know too much about the subject. There are a couple of services that can be employed in order to transform the process of turning from a newbie into an experienced investor into a smooth one. Take this chance and make it work!

Critics state that using Bitcoins is unsafe because

- They have no authentic value

- They are not regulated

- They can be used to make illegal transactions

Reasons why it is worth using Bitcoins.

1. Quick payments - When payments are made by using banks, the transaction takes some days, similarly wire transfers also take a long time. On the other hand, virtual currency Bitcoin transactions are generally more rapid.

"Zero-confirmation" transactions are instantaneous, where the merchant accepts the risk, which is still not approved by Bitcoin block-chain. If the merchant needs an approval, then the transaction takes 10 minutes. This is much more rapid than any inter-banking transfer.

2. Inexpensive - Credit or debit card transactions are instant, but you are charged a fee for using this privilege. In the Bitcoin transactions, the fees are usually low, and in some cases, it is free.

3. No one can take it away - Bitcoin is decentralized, so no central authority can take away percentage from your deposits.

4. No chargeback - Once you trade Bitcoins, they are gone. You cannot reclaim them without the recipient's consent. Thus, it becomes difficult to commit the chargeback fraud, which is often experienced by people with credit cards. People purchase goods and if they find it defective, they contact credit cards agency to make a chargeback, effectively reversing the transaction. The credit card company does it and charges you with costly chargeback fee ranging from $5-$15.

5. Safe personal details - Credit card numbers get stolen during online payments. A Bitcoin transaction does not need any personal details. You will need to combine your private key and the Bitcoin key together to do a transaction.

6. Eliminates fraud risk - Only the Bitcoin owner can send payment to the intended recipient, who is the only one who can receive it. The network knows the transfer has occurred and transactions are validated; they cannot be challenged or taken back. This is big for online merchants who are often subject to credit card processors' assessments of whether or not a transaction is fraudulent, or businesses that pay the high price of credit card chargebacks.

7. Data is secure - As we have seen with recent hacks on national retailers' payment processing systems, the Internet is not always a secure place for private data. With Bitcoin, users do not give up private information.

a. They have two keys - a public key that serves as the bitcoin address and a private key with personal data.

b. Transactions are "signed" digitally by combining the public and private keys; a mathematical function is applied and a certificate is generated proving the user initiated the transaction. Digital signatures are unique to each transaction and cannot be re-used.

c. The merchant/recipient never sees your secret information (name, number, physical address) so it's somewhat anonymous but it is traceable (to the bitcoin address on the public key).

8. Convenient payment system - Merchants can use Bitcoin entirely as a payment system; they do not have to hold any Bitcoin currency since Bitcoin can be converted to dollars. Consumers or merchants can trade in and out of Bitcoin and other currencies at any time.

9. International payments - Bitcoin is used around the world; e-commerce merchants and service providers can easily accept international payments, which open up new potential marketplaces for them.

10. Easy to track - The network tracks and permanently logs every transaction in the Bitcoin block chain (the database). In the case of possible wrongdoing, it is easier for law enforcement officials to trace these transactions.

11. Micropayments are possible - Bitcoins can be divided down to one one-hundred-millionth, so running small payments of a dollar or less becomes a free or near-free transaction. This could be a real boon for convenience stores, coffee shops, and subscription-based websites (videos, publications).

12. No Taxation - When you make purchases via dollars, euros or any other government fiat currency, you have to pay an addition sum of money to the government as tax. Every purchasable item has its own designated tax rate. However, when you're making a purchase through Bitcoin, sales taxes are not added to your purchase. This is deemed as a legal form of tax evasion and is one of the major advantages of being a Bitcoin user.

With zero tax rates, Bitcoin can come in handy especially when purchasing luxury items that are exclusive to a foreign land. Such items, more often than not, are heavily taxed by the government.

13. Flexible Online Payments - Bitcoin is an online payment system and just like any other such system, the users of Bitcoin have the luxury of paying for their coins from any corner of the world that has an internet connection. This means that you could be lying on your bed and purchasing coins instead of taking the pain of travelling to a specific bank or store to get your work done.

Moreover, an online payment via Bitcoin does not require you to fill in details about your personal information. Hence, processing Bitcoin transactions is a lot simpler than those carried out through U.S. Bank accounts and credit cards.

14. Concealed User Identity- All Bitcoin transactions are discrete, or in other words Bitcoin gives you the option of User anonymity. Bitcoins are similar to cash only purchases in the sense that your transactions can never be tracked back to you and these purchases are never connected with your personal identity. As a matter of fact, the Bitcoin address that is created for user purchases is never the same for two different transactions.

If you want to, you do have the option of voluntarily revealing and publishing your Bitcoin transactions but in most cases users keep their identities secret.

15. No outside interventions - One of the greatest advantages of Bitcoin is that it eliminates third party interruptions. This means that governments, banks and other financial intermediaries have no authority whatsoever to disrupt user transactions or freeze a Bitcoin account. As mentioned before, Bitcoin is based strictly on a peer to peer system. Hence, the users of Bitcoin enjoy greater liberty when making purchases with Bitcoins than they do when using conventional national currencies.

c. Where can you find bitcoin prices?

Regardless whether you're a bitcoin holder or planning to get some bitcoin, you will always want to know how much such cryptocurrencies value are to the currency choice of yours.

But what is Bitcoin's current price?

The best place to help you find out the latest price of bitcoin is the exchange. Some of the Bitcoin's exchanges are:

- Bitstamp
- Bitfinex
- BTC-e
- Coindesk

- coincheckJPY
- bitflyerJPY
- coinbaseUSD
- bitstampUSD
- zaifJPY
- fiscoJPY
- krakenEUR
- korbitKRW
- hitbtcUSD
- lakeUSD
- btcboxJPY
- krakenUSD
- wexUSD
- bitbayPLN
- itbitUSD
- coinsbankRUB
- zyadoEUR
- cexUSD
- coinbaseEUR
- coinsbankGBP

Knowing the Bitcoin's current price is one thing. However, pretty soon you'll be interested in knowing where the prices will go in the future. This requires sound forecasting approaches. Forecasting the Bitcoin price movements of anything that is traded at an exchange can be a risky undertaking—nobody or institution can be right all the time. As a matter of fact, the majority of bitcoin traders have lost lots of money, into such attempts.

At present, there are two main approaches to forecasting price movements: fundamental analysis and technical analysis. While the fundamental analysis approach examines the underlying forces of the cryptocurrency economy, the company or any security, the technical analysis tries to predict the direction of bitcoin prices based on historical market data, and volumes found on the price charts.

I know you may be wondering, "Where can I find bitcoin price charts?"

To conduct technical analysis on the bitcoin price and volume history, you must use the bitcoin price charts that display data in a more readable manner than just the plain number tables.

You can find the bitcoin prices charts from the following popular sites:

- Bitcoinity
- Zeroblock
- Candlestick charts
- Bitcoin Wisdom
- TradeBlock

Performing technical analysis on Bitcoin Price

To perform any technical analysis or predict the movement of bitcoin prices in reference to bitcoin price history, you'll need bitcoin charts to start.

Two type of chart is usually available for reference.

Simple Price chart

A simple price chart which provide the closed price. User can shorten the gap by selecting respective chart view based on a period of time from a day, a week, a month or a year.

This method provide the user a quick summary of the bitcoin prices have been doing in relevance to their viewing period but such chart lack data for pro-user who demands more data.

Candlestick chart

this type of chart is usually adapted and used within the trading industry(commodities, forex, etc.) as it provide more data that just the closing price. Each candle shows the opening price, the lowest and highest within the selected time frame as well as the closing price for the day.

Besides providing such detailed information to its user, it also provide a colored candle indicating if the closing price was higher or lower than the opening price. This type of candle is usually represented by two color or two type. Bullish is usually generated if there's a positive growth within the comparison timing or Bearish is usually generated if there's a negative growth within the comparison timing.

Chapter 4:

How to Buy and Sell Your Bitcoin

a. Should you invest in bitcoin?

F irst, let's define some terms. By "invest," do you mean purchasing a large amount of BTC hoping you can sell it for a massive profit later (speculating)? Or do you mean buying and selling BTC to take profit from short-term price fluctuations (trading)?

If you mean either of these things, then:

Short Answer: Probably not.

Long Answer: This is not a book of investment advice. This is an introductory text who's intended audience is people who know little to nothing about Bitcoin. If there is one piece of investment wisdom I CAN provide, it is to not invest in things you do not understand. This book does not contain enough technical information for you to use as a basis for investment decisions.

Haven't people made a lot of money investing in Bitcoin?

Yes, they have. In early 2011, one BTC was worth $1 USD. Today, in 2017, one BTC is $2100. If you had invested in BTC when it was below one dollar, you would have amassed a life-changing amount of money in less than a decade. Plenty of people did. Some were developers or extremely knowledgeable hobbyists who saw the potential in Bitcoin and made a decision that turned out to be a mind-bogglingly profitable. Others were folks who heard about Bitcoin and, knowing little more about it beyond where to buy some, bought a bunch because other people were doing it. The first group of people was investors who knew what they were doing and took a calculated risk. The second group was gamblers who got lucky.

Both groups made money, right?

Sure. Bitcoin has made millionaires.

It also ruined people. The rise from cents to thousands of dollars was not a straight line. From late 2013 to early 2014, BTC experienced a drop from over $1200 to around $350. It did not return to the $1200 level until March of 2017. People who bought in at the top and sold in frustration when the price dropped lost a lot of money. At one point in 2014, the top post on the popular Bitcoin forum on Reddit was a link to a suicide hotline. It was not a joke.

If I may offer another piece of investment advice in addition to "Don't invest in things you don't understand," it would be "Do not invest money you can't afford to lose."

There are two things that must be true before you invest any significant amount of money into BTC: 1) You must be familiar with the technology at a level beyond what is presented in this book. 2) You must already be a successful investor or trader in commodities, stocks, or currencies.

If you are a new speculator or trader, Bitcoin is not a place for you to learn. If you are an experienced speculator or trader, you already know better than to jump into something blind.

For the record, I do not believe Bitcoin has exhausted its upward potential. There is still profit to be made in the long term, but only by those who tread carefully and put in the effort to learn what they're investing in. What I do recommend for Bitcoin newbies is learn the process of obtaining, spending, and securely storing Bitcoin by actually doing it yourself. Instead of money, invest time and energy. See how it works. See what problems Bitcoin solves and what problems it still has to overcome before it can achieve mass adoption. Participate in (or at least observe) discussions on places like Reddit or bitcoin talk. It won't take long for you to appreciate Bitcoin's potential AND its weaknesses, enabling you to make investment decisions properly.

b. How bitcoin exchange operates?

To help you understand how Bitcoin exchanges operate, let's first explain how banks or bank-like services operate in a classical economy. You offer the bank some money — which can be a deposit — and the bank promises to provide you with that money later. Obviously, the bank doesn't just take your cash and place it in a box somewhere in the back room. All that the bank does is promise you next time you show up for the money you'll get it back.

Chapter 4: How to Buy and Sell Your Bitcoin

The bank will take the money and invest or put it somewhere else. In other words, the bank will probably store some money as a reserve to allow it pay out the demand for withdrawals which it will be facing on a typical day. As a matter of fact, the majority of banks use a fractional reserve where they maintain a certain fraction of all the demand deposits on the reserve just in case a potential client shows up to withdraw.

What about the bitcoin exchanges?

Bitcoin exchanges are businesses that function (at least from a user interface point of view) as a bank. They accept deposits of inform of bitcoins and will—just like conventional banks—promise to provide them back when demanded later. You can also transfer the fiat currency such as Dollars and Euros into the Bitcoin exchange by transferring from your bank account.

The exchange promises to pay you back either in fiat currency or bitcoins or even both. The exchange can let you perform various banking-like transactions such as making and receiving Bitcoin payments and exchanging Bitcoins for fiat currencies. Typically they will do this by finding some client who wants to buy your bitcoins with fiat currencies and some other client who wants to sell the bitcoins for the fiat currencies and match them up.

In other words, the Bitcoin exchange will find the clients who are willing to take opposite directions in a cryptocurrency transaction. If there is a mutual agreement about the prices, the clients will proceed to consummate the transaction.

Let me use an example to drive the point home.

Suppose your bank account holds $5000 and 2BTC and you use the exchange. You put in an order to buy 3BTC for $700 each. Now the exchange finds someone who is willing to sell Bitcoins at that price, then the transaction occurs. You'll end up with 5BTC in your account and $2100 will be deducted from your bank account to buy the Bitcoins.

The vital thing to note about this transaction is that it has occurred between me and the Bitcoin seller on the same exchange. At this moment, no transaction has happened on the Bitcoin blockchain. The exchange doesn't require to go to the blockchain to transfer the bitcoins or dollars from one account to the other. All that occurs in this transaction is that the exchange will now be making a different promise to you than it was making before.

Before it said, "I'll give you 5000 USD and 2BTC." With the transaction having taken place, it will now promise you with "I'll give you $2900 and 5BTC" which is just a change in promise. Modern bitcoin exchanges are P2P cryptocurrency platforms that allow users to swiftly and easily convert their bitcoins into the fiat currencies and vice versa. Unlike other traditional banks, these exchanges use the next-generation blockchain platform that allows any person—whether buyers, sellers or merchants—to buy and sell bitcoins in the most reliable and convenient way.

c. Step-by-step guide on how to buy bitcoin online

The first step towards getting started with Bitcoins is setting up the wallet. The Bitcoin wallet is just like your normal wallet that keeps notes and coins. The Bitcoin wallet allows you to store your Bitcoins. In particular, the Bitcoin wallet is a digital file created by the Bitcoin software that creates and stores Bitcoins.

Being a decentralized system, the Bitcoin wallet can be stored on multiple computers. The wallet file comprises of two main parts namely: the file—which stores your bitcoins—and the wallet application—the program that opens the file on your computer. The Bitcoin wallet has two parts: the wallet file, that stores the Bitcoins, and the wallet software—sometimes called the Bitcoin client—that opens the wallet files.

It is possible to store your wallet file on a flash disk and open it on any computer so long as it has the wallet application installed. There are multiple types of wallets that you can choose from. They vary in features and the devices on which they are installed. The easiest solution for getting meaningful amounts of Bitcoins is to exchange your currency for Bitcoins. You can use any of the bitcoin exchanges that I have mentioned above to help you convert fiat currency into bitcoins.

Basically, there are two ways to buy Bitcoin online: the fastest way and the cheapest way. The fastest way to buy Bitcoin is to use your credit card. This process takes can take a few minutes but you'll incur higher fees. A cheaper way, though the considerably more involved approach is to purchase bitcoins on an exchange. Let's dive in to explore these methods.

How to buy bitcoins using your credit card

Multiple wallets have partnered with Simplex—an online payment system—that makes it easy for buyers to buy bitcoin using their credit card. Follow these steps to help you securely buy your bitcoins:

- Go to the bitcoin exchange platform and locate the link "Buy Bitcoin" or whatever link that is almost similar to "Buy Bitcoins."

- Enter your Bitcoin wallet address and select the desired bitcoins you want to buy.

- Explore the "Buy Bitcoin" page and input the number of bitcoins you want to buy.

- Now, specify the Bitcoin address where the purchased Bitcoins will be sent. To achieve this, open your Bitcoin wallet, copy the Bitcoin address and paste it into "Bitcoin address" field. The bitcoins will now be sent to this address. Recall that you can always download the Bitcoin wallet address for free.

- Enter the billing information. Once you have entered the Bitcoin address and chosen the purchase amount, click on the "Continue" button or any button similar. From here you'll now be re-directed to the Simplex's checkout form. Follow the steps, enter your information, and the bitcoins will be sent to your address within minutes!

Next up, let's now explore how to buy bitcoins on an exchange.

First and foremost, you will need to select an exchange based on your country and the desired purchase currency. Here are steps that can help you buy bitcoins on a bitcoin exchange:

- Choose an exchange platform by entering the country of residence and chose from the list of approved exchanges.

- Setup an account on the bitcoin exchange platform. Depending on the exchange platform, various forms of identification may be required.

- Link up one or more bank accounts to the exchange account. Any purchases or sells that you make will be deducted from and deposited to the selected bank account.

- Select the number of bitcoins to buy. Navigate to the bitcoin exchanges "Buy feature" and: select the number of bitcoins to buy, enter your bitcoin wallet address, confirm the purchase, and wait for the purchased Bitcoin transaction to be completed.

d. Alternative site for buying and selling – P2P bitcoin exchange

Simply because a website announces that it can buy and sell bitcoins isn't a surefire way to determine whether it's real or fake. Many of the websites that claim to buy and sell bitcoins are rather recent, thus, have little information to get the word out of their products and services.

Obviously, the first thing that you consider when figuring out the top bitcoin exchanges is how safe they are. Ultimately, it will boil down to asking, "is it a trustworthy P2P exchange ensuring transparency in the transactions?" It's also important to lookout for the currency pairs are that are available. For instance, are you looking to trade your bitcoins for USD or Euros?

Besides Coindesk, here are other P2P sites where you can buy and sell bitcoins:

- Coinbase
- BitStamp
- eToro
- Kraken.com
- Bittrex
- LocalBitcoins
- Yobit
- Blockchain
- HitBTC
- BitHump
- Poloniex
- Bitso
- Bitpay
- Coincheck
- Zebpay
- Bitfinex
- Paxful
- Yobit
- BitMex

- Gemini
- BitFlyer
- Remitano
- BTCMarkets

e. Buying and selling bitcoin through ATM and offline

Buying through ATM

Another way to buy Bitcoin that is becoming increasingly popular is through Bitcoin ATMs. These devices are cropping up all over the place, from malls to airports to city centers. They look a lot like traditional ATM's, but there are some important differences. Primarily, Bitcoin ATM's don't connect to any banks. They are connected, through the Internet, only to the universe of the Bitcoin network. Many Bitcoin ATM's allow for bidirectional exchange, meaning that you can either insert cash to be converted to Bitcoin and transferred to a public key address, or you can have Bitcoins from your own account converted into cash and dispensed by the machine. Some only handle transfers one-way or the other.

ATM's can provide a more anonymous way to buy into Bitcoin without syncing your bank account to a platform like Coinbase. However, there may be high transaction fees and limits on how much you can deposit or withdraw depending on the machine. One way that you can search for Bitcoin ATM's in your area is by using the Coin ATM Radar website at https://coinatmradar.com

Getting Bitcoin From Someone Else

Since the very beginning of Bitcoin, one of the most common ways to get started the currency has been to find someone willing to gift or sell some to you. As Bitcoin is a digital asset, it should not be too surprising that much of the Bitcoin community exists online. Forums such as bitcointalk.org or Reddit's r/bitcoin are good places to engage with other Bitcoin users. Some Bitcoin enthusiasts are happy to donate a small amount to a newcomer to help them establish their first wallet. After all, the more people that use Bitcoin, the higher the demand, thus the more valuable it will become, at least in theory. Through that lens, it makes sense from a long-term financial perspective to help new users get started even if means spending a little of your own coin initially.

There are also a variety tools online to find local Bitcoin exchanges, where people actually meet up offline, in person, to trade with Bitcoin. This can be a great way to avoid transaction fees, meet fellow Bitcoin enthusiasts in your area, and potentially increase the level of transaction anonymity. As with any scenario that involves meeting someone "from the Internet," use your judgment if meeting up to exchange Bitcoin locally.

Chapter 5 :

Bitcoin Wallet

a. What is bitcoin wallet

Before you choose to make your first transactions using bitcoin take a little time to educate yourself and arm yourself with the correct tools to keep you safe.

Securing your bitcoin wallet is a process similar to choosing a bank account that suits your needs. As with all choices that affect your financial safety your personal needs are the most important points to consider.

Consider the device you will use when controlling your bitcoin purchases and accounts. There are some great apps available to use with your iPhone or Android cell if that is your choice of device. Below we will explore some different options that may be perfect for you.

Copay has a simple easy to use interface that makes it a popular choice for new users of bitcoin. Available on Linus, Max OS X, Android phone and Windows it is a great choice for use across multiple platforms. Business users will often opt for Copay as they have a shared account feature. This means that multiple users are required to sign for each transaction and ensures a higher level of security.

Airbitz is the choice for many less techy minded users! The account is programmed to manage all your usernames and passwords without infiltrating your actual funds. Easy to use and perfect for the beginner another current feature of this account is discounted goods at Starbucks!

b. Bitcoin private key

A Bitcoin private key can be described as a secret number created to let people spend their Bitcoins or make irreversible transactions. Users are given a Bitcoin private key when they are issued with a Bitcoin address. Typically, it has a 256-bit number and may be used to sell, accept, donate Bitcoin, hence must be kept really safe. For instance, a Bitcoin private key may look like this: 18Qs4IuA5d5ViEiPWYau6fhRTHEFZ9XaLo

Keeping Your Private Keys Safe

In the past, several secret/private keys or backup seed have been lost due to the storage medium on which they are saved. Frequently used mediums of storing private keys are listed below with some of their weaknesses.

1. Storage On A Piece Of Paper

Whether the information is written, printed or laminated, a number of things could go wrong with the storage medium, and they are not limited to the following:

- The paper may be discovered and stolen

- The paper could be torn, burnt, spoiled, or damaged by smoke

- A hand-written paper might not be legible; laminated paper is susceptible to being ruined while attempts to print on paper could be unsuccessful if the paper is wet.

2. Storage On A Flash Drive

- The possibility of breakage exists

- It can be affected by fast changing magnetic fields, for example, MRIs

- They may be affected by fire and smoke

- Many of these drives are not designed for storing things in the long run

- Can become corroded from salt water or some atmospheric conditions

- You may find it hard to retrieve your data from it

- It can be adversely affected by harsh environmental factors

- In general, flash drives aren't recommended for long-term storage

3. Storage In The Cloud

- There is a risk of hackers attempting to steal the private keys

- Other people may have access to your cloud storage and take the keys

4. Storage On A Computer

- They are susceptible to crashes which make data recovery costly

- Computers are prone to physical attacks and may get burnt or damaged by smoke

- The data on conventional hard disc drive may be degraded by strong magnetic fields and could get destroyed physically.

- Mishaps might occur that will bring about data loss

- It is ill-advised to store up data meant to last for long on Solid state drives (SSDs) if they are not going to be powered.

- If the computer is linked to the internet, it is prone to attacks from hackers who might want to break into it, to steal the key irrespective of the encryption technology employed.

- The use of a computer for storage of private keys is often associated with a broad range of threats like firmware exploits, the use of malicious USB cords and 0-day exploits.

- The use external hard disk drives for storage are limited to just couple of years as a minimum if stored appropriately

- If computer is not linked to the net, the safety it provides is function of the encryption technology used and doesn't negate the fact that an individual may still enter the location illegally and copy the data with no one taking notice

5. **Storage On Digital Media Like Cd, Floppy Disk, Laserdisc, Or Mini-Disc**

- There is a high tendency for plastics to stop working after a while.

- Exposure to adverse environmental conditions such as heat, humidity, regular light, all kind of chemicals, and the oxygen in the air may degrade them. It could also result in data loss when private keys are stored on a medium derived from plastic or written/printed on plastic.

- Plastics could get burnt or become damaged by smoke

- The risk of bodily harm occurring exist, thereby making it not viable or costly to recover the lost data

- There is a probability that magnetic media such as tapes and floppy disc could be damaged by magnetic fields

c. Type of bitcoin wallet

In the Bitcoin ecosystem, Bitcoin wallets are one of the most actively developed applications. Choosing a wallet depends on the use and user expertise, and it is highly subjective. Moving money between Bitcoin wallets is fast, cheap and easy. So, try several different wallets until you find one that fits your needs.

First, you need to get a Bitcoin wallet.

According to the platform, you can choose from a variety of Bitcoin wallets:

- Desktop wallet

Created as a reference implementation, the first type of Bitcoin wallet was a desktop wallet. Many users like the desktop wallet for the autonomy, control, and features they offer. They can run on Mac and Windows but vulnerable because of these platforms are poorly configured and insecure.

- Mobile wallet

A mobile wallet is the most common type of Bitcoin wallet. These wallets are a great choice for new users because they run on smartphone operating systems like Android and Apple iOS.

- Cloud wallet

Through a web browser, Cloud wallets are accessed and store the user's wallet on a server operated by a third party. It relies entirely on a third-party server and similar to webmail. The advantage of Cloud-based wallets is that you can access them from anywhere with any device. The major disadvantage is if you can operate it correctly, you can place the third party (the web page authority) in charge of your private keys. Using Coinbase, users in the Europe and US can buy Bitcoin through its exchange services. The blockchain is also a popular web-based wallet.

- Hardware wallet

Hardware wallets are operated through near-field communication (NFC) on a mobile device or via USB with a desktop web browser. These wallets are considered very secure and suitable for storing large amounts of Bitcoin.

- Ledger USB wallet

The Ledger USB wallet is available for a reasonable price and uses smart card security.

d. Paper/offline wallet and its advantage

All the Bitcoin wallets can be classified as either online (hot) or offline (cold). Those wallets that are connected to the internet are called the hot wallets while those that are offline are called cold wallets. The paper/offline wallets are still the most popular and cheapest options for keeping bitcoins safe today.

There are multiple websites that offer paper bitcoin wallet services. These sites generate a bitcoin address for your bitcoins storage and create an image with two QR codes:

- The public address that you'll use to receive bitcoins

- The private key, that you will use to spend the bitcoins stored at that address.

These wallets are considered to be more secure—and can be relied upon to store larger amounts of bitcoins—compared to hot bitcoin wallets that are suitable for storing frequently accessible bitcoin funds. These wallets are safe and hacker-proof because whenever you make a transaction, they'll ask you to confirm each transaction by pressing a button on the crypto device. They are much hacker-proof. At least I've never heard of any being hacked before.

There are different types of offline wallets namely: Trezor.io, and KeepKey and Ledger Nano S. Each of these wallets have different features, therefore, you must do due diligence and research before you settle on one. For instance, the Trezor.io provides an efficient customer service and has an outstanding communication team that supports clients.

Here are steps of generating a paper wallet on BitAddress.org:

- Open the BitAddress.org in your browser. The BitAddress will ask you to generate some random number by either randomly typing the characters into the form or moving the cursor around.

- You will then be presented with your public and private keys together with their respective QR codes. Don't scan these codes

- Now click the "Paper wallet" tab.

- Choose the number of addresses that you want to be generated. If you don't want to keep the bitcoin artwork, simply click on the "Hide art?" button.

- Click the on the "'Generate" button to generate the new wallets.

- Once the wallets have been generated, click on the "Print'" button to create a hard copy wallet.

- Your browser will then ask you to choose the printer that you wish to use. If you're using Google Chrome, you can also save the page in form of a PDF file.

- Take note of its public addresses, or you can scan the public address QR code in your bitcoin application and start depositing the bitcoins

e. Bitcoin wallet balance

If you are unsure whether your wallet address has received funds you don't have your wallet at hand, you can still verify the balance by using a Blockchain Explorer. This is because all the all the bitcoin addresses are visible publicly on the bitcoin blockchain and you can look for them up if you know the address or the transaction identity.

The Transaction details on the blockchain will list all the inputs as well as the outputs which have been used by the transaction. Clicking on any of the transaction details will show you the bitcoin address and the transaction when you choose the Hash/Transaction ID) in the BlockExplorer on Blockchain Explorer. Looking at the transaction

ID using the block explorer you'll see that your transaction is received but can either be confirmed or unconfirmed by the network.

The time it for the first confirmation to take place will depend on the fee that the transaction is providing to the miner. The higher the transaction fee, the higher the likelihood that it will be included in the upcoming block.

You can also use alternative tools to help you confirm your wallet balance. One such tool is the BitRef. BitRef is an online tool that you can easily integrate with Google Chrome, Mozilla or Opera to check the bitcoin address balance. What you need for such a task is a device with the Internet connection and any valid Bitcoin address string.

f. Things to look out for when using your bitcoin wallet

In recent times, numerous threats to the security of stored data on wallets during execution of online transactions have been reported. With the bitcoin cryptocurrency gaining popularity as a form of payment, you should be aware of the security threats involved and how to secure your bitcoin wallet.

Below are tips that can help you to protect your Bitcoin wallet:

#1: *Always use a versatile Bitcoin client*

For the purpose of anonymity and privacy, hiding your IP address is paramount. Therefore using Bitcoin client that can change your address with every transaction you perform can help you secure your wallet.

Likewise, you can separate the Bitcoin transactions into different wallets based on their importance. The best practice is: keep a Bitcoin wallet for day-to-day transactions of small amounts only, and only top it when necessary.

#2: *Always protect your identity*

Be careful when sharing any data about your Bitcoin transactions in the public spaces such as the internet to avoid revealing your personal identity details to malicious users.

#3: *Consider using an escrow service*

When you want to buy or even sell anything and you're not sure who is on the other end of the web, use the escrow service. With escrow service, the person on the other end will send Bitcoins to the escrow service that will guarantee that a transaction has been finalized before releasing payment.

#4: Always backup your Bitcoin wallet

It is critically important to have a backup policy. You should always make frequent updates, make different media and locations back-ups.

#5: Consider encrypting your wallet

Encrypting your Bitcoin wallet is crucial when your wallet is stored online. Using a strong password is equally important. Tools such as DESlock+ can help you encrypt your Bitcoin wallet.

#6: Consider the two-factor authentication

Always undertake an extensive selection process that determines a truly reliable authentication service. The best practice is to use the two-factor authentication process wherever possible to provide security to online wallets.

#7: Don't use wallets on mobile devices

As a rule of thumb, don't use your Bitcoin wallet on mobile devices especially when transacting huge sums of money. The mobile devices can get lost or compromised where you'll lose your Bitcoins.

#8: Consider using the multi-signature addresses

For corporate transactions,--or transactions that demand a high level of security— you can use multi-signature addresses that use more than one key for encryption and decryption of data.

#9: Always update your systems on a regular basis

Any application can have flaws, so it is important to constantly update your Bitcoin software and the operating system. The

Chapter 5 : Bitcoin Wallet

Bitcoin wallet can be affected by any malware that's hosted on your hardware.

Chapter 6 :

Bitcoin Mining in Theory

a. Bitcoin economics and currency creation, Independent verification of transactions

If Bitcoins come from mining, it might be quite tempting to conclude that one should drop everything and become a Bitcoin miner. One of the first things that newcomers to the Bitcoin space often hear about is mining. At a glance, this can look like a primrose path to "free money." It can sound like mining is as easy as firing up an application your laptop and watching the Bitcoins come rolling in! Unfortunately, like most things involving "free money," the reality is that mining is not so simple.

Earlier we looked at the blockchain and how transactions are stored throughout a distributed network of computers. We know that each transaction is verified and added the blockchain, but how exactly does this happen?

This is where miners come in. Bitcoin miners use special software to solve complex math problems that are used to verify transactions, maintain the blockchain and add blocks. By checking a new transaction against the public ledger of previous transactions (the blockchain), a "node" (a particular mining station) is able to distinguish between a valid transaction and an invalid one.

If someone attempts to spend Bitcoins that don't exist the system will say, "Hey, wait a minute, this doesn't match up with the history on the blockchain..." and the transaction will be rejected. Miners handle the heavy duty computer processing that it takes to check all new Bitcoin transactions, verify them, and add them to the blockchain.

In exchange for solving blocks, miners are rewarded with a certain amount of new Bitcoin, thus adding a little bit at a time to the global volume of available Bitcoins in circulation. This incentive encourages more people to mine, leading to a more secure system through wider distribution.

In the early days of Bitcoin, mining was something that could be done on pretty much any old computer and the reward for "discovering" a block was ample, while the overall value of Bitcoin was extremely low and there were not that many Bitcoins in circulation. Over time, as more and more people began to use Bitcoin, and also to become miners, the conditions changed.

Built into the protocol behind Bitcoin is a relationship between the number of miners and the level of difficulty involved in solving the problem, or "mining" each block. In theory, we can imagine that if more miners enter the scene, more blocks will be created at a faster pace. The Bitcoin protocol works in such a way that as more blocks are created, the rate of difficulty involved in solving the complex math problems required to successfully "mine" a block goes up. By making it harder to mine a block, the rate of block creation goes down. This relationship between the amount of Bitcoins that exist and the level of difficulty involved in mining new ones keeps the ecosystem stable over time.

The level of difficulty, today, required to mine a block is so resource-heavy, both in computer power and electricity, that it requires special equipment. In most cases, the cost of a mining operation far outweighs what one could hope to earn from mining Bitcoin for a very long time. Many miners today operate in collectives known as "pools," where members combine resources and split rewards. Joining a mining pool is one way to increase the odds of recouping the costs of mining equipment and potentially making a profit. It is still possible to earn Bitcoin today through mining, but it is definitely not easy or free to get started.

b. Mining nodes

The mining nodes are those computers that compete to develop new blocks by executing a specialized hardware to solve the PoW (proof-of-work) algorithm. Some mining nodes can also be full nodes, maintaining the full copy of the blockchain database, while others may be lightweight nodes participating in pool mining.

Difficulty factors

The main reason behind the development of Bitcoin was to permit users to conduct P2P transactions over the web without going through middlemen or other trusted third parties such as banking and financial organizations. To fulfill this noble objective, the Bitcoin protocol has to meet specific criteria. Apart from being fungible, scarce, and highly divisible, it is crucial for the Bitcoin network to maintain a level of consistency. The consistency of the Bitcoin network takes a special place because of its decentralized nature.

We all know that the Bitcoin system is a decentralized, open source, digital currency where miners—who are integral within the Bitcoin community— contribute their processing power to unearth new blocks and confirm the transactions over the ecosystem. This is performed by comparing the transaction identities with the existing records on the Blockchain and placing the new transaction records into fresh blocks. Technically speaking, the time taken by any Bitcoin miner or a mining pool to unearth a new block is directly proportional to the sum processing power of the entire network.

In technical terms, this is called the hash rate. Today, Bitcoin mining is carried out using specialized circuits known as Application Specific Integrated Circuits (ASICs) which are customized for mining. These circuits have components that consume a lot of processing power. In other words, their hash rate must be high enough to validate transactions on the Blockchain.

As technology advances, superior ASICs are being conceived and fabricated that are far more efficient, both in terms of power consumption and hash rate. With the new ASICs replacing the older versions, the total hash rate has skyrocketed in the Bitcoin ecosystem. Ideally, it would imply that the total time required to generate a new block drastically reduces too, leading to a corresponding increase in the payouts. Now, to prevent such a scenario, the difficulty level of the Bitcoin network is always exploited to keep the total time taken for new block discovery constant.

The Bitcoin network automatically alters the difficulty level for Bitcoin mining to ensure the discovery of new blocks every 10 minutes by miners. To be more specific, the mining difficulty is influenced by two factors. First and foremost, there is a global block difficulty. The global block difficulty forces the valid blocks to have a hash that is below the target.

Secondly, the number of users actively participating in the mining process.

In Bitcoin universe, the mining difficulty automatically adjusts after every 2016 blocks on the Bitcoin network. Now, depending on how many users were actively mining – together with their combined hashpower— and the time it takes to find the 2016 blocks, the difficulty can either go up or down. As the mining difficulty increases, miners should acquire more powerful hardware to accommodate the adjustment. And this is why the CPU, FPGA and GPU mining were rendered obsolete when ASIC hardware was developed by manufacturers.

It is also vital to note that there is no maximum mining difficulty for the bitcoin network. There is a possibility that the difficulty will continue to rise until all the bitcoins are mined. This is expected to happen by the year 2140. Also, the mining difficulty factor can assume on large leaps and bounds compared to the preceding 2016 blocks. An increase in the difficulty by more than 15% isn't uncommon in the bitcoin universe, yet it makes a huge difference for miners who are unwilling to upgrade their hardware.

Although bitcoin mining difficulty can also go down, it is unlikely that this will ever happen more than twice in any one year. In fact, when you look at the most recent charts, you'll realize that the mining difficulty has a tendency to continue increasing over time. As of writing this, there are only five adjustments where the difficulty has gone down since July 2015.

c. Mining the bitcoin block

i. Proof of work algorithm

At the very heart of the operations in a blockchain is this concept. Proof of work was an important part of the original blockchain role as the transaction authenticator. Proof of work is what provides the right to take part in the blockchain and it is displayed as a large hurdle that stops users from making changes to records stored on the chain without providing a new proof of work.

It is one of the main building blocks simply because it can never be undone and is cryptographically secured through the hashes that are used to prove its authenticity. However, it an expensive concept to maintain, estimated to cost about $600 million a year jut for bitcoin and that means there could be future issues of scalability and security. This is because it depends entirely on the incentives for the miners – mining will decline as time goes by. A better solution is called "proof of stake" – much cheaper for enforcing but way more expensive and much more difficult to compromise. This concept will determine who is allowed to update the consensus and stop the underlying blockchain from being forked.

ii. Difficulty factors

The main reason behind the development of Bitcoin was to permit users to conduct P2P transactions over the web without going through middlemen or other trusted third parties such as banking and financial organizations. To fulfill this noble objective, the Bitcoin protocol has to meet specific criteria. Apart from being fungible, scarce, and highly divisible, it is crucial for the Bitcoin network to maintain a level of consistency. The consistency of the Bitcoin network takes a special place because of its decentralized nature.

We all know that the Bitcoin system is a decentralized, open source, digital currency where miners—who are integral within the Bitcoin community— contribute their processing power to unearth new blocks and confirm the transactions over the ecosystem. This is performed by comparing the transaction identities with the existing records on the Blockchain and placing the new transaction records into fresh blocks. Technically speaking, the time taken by any Bitcoin miner or a mining pool to unearth a new block is directly proportional to the sum processing power of the entire network.

In technical terms, this is called the hash rate. Today, Bitcoin mining is carried out using specialized circuits known as Application Specific Integrated Circuits (ASICs) which are customized for mining. These circuits have components that consume a lot of processing power. In other words, their hash rate must be high enough to validate transactions on the Blockchain.

Chapter 6 : Bitcoin Mining in Theory

As technology advances, superior ASICs are being conceived and fabricated that are far more efficient, both in terms of power consumption and hash rate. With the new ASICs replacing the older versions, the total hash rate has skyrocketed in the Bitcoin ecosystem. Ideally, it would imply that the total time required to generate a new block drastically reduces too, leading to a corresponding increase in the payouts. Now, to prevent such a scenario, the difficulty level of the Bitcoin network is always exploited to keep the total time taken for new block discovery constant.

The Bitcoin network automatically alters the difficulty level for Bitcoin mining to ensure the discovery of new blocks every 10 minutes by miners. To be more specific, the mining difficulty is influenced by two factors. First and foremost, there is a global block difficulty. The global block difficulty forces the valid blocks to have a hash that is below the target. Secondly, the number of users actively participating in the mining process.

In Bitcoin universe, the mining difficulty automatically adjusts after every 2016 blocks on the Bitcoin network. Now, depending on how many users were actively mining – together with their combined hashpower— and the time it takes to find the 2016 blocks, the difficulty can either go up or down. As the mining difficulty increases, miners should acquire more powerful hardware to accommodate the adjustment. And this is why the CPU, FPGA and GPU mining were rendered obsolete when ASIC hardware was developed by manufacturers.

It is also vital to note that there is no maximum mining difficulty for the bitcoin network. There is a possibility that the difficulty will continue to rise until all the bitcoins are mined. This is expected to happen by the year 2140. Also, the mining difficulty factor can assume on large leaps and bounds compared to the preceding 2016 blocks. An increase in the difficulty by more than 15% isn't uncommon in the bitcoin universe, yet it makes a huge difference for miners who are unwilling to upgrade their hardware.

Although bitcoin mining difficulty can also go down, it is unlikely that this will ever happen more than twice in any one year. In fact, when you look at the most recent charts, you'll realize that the mining difficulty has a tendency to continue increasing over time. As of writing this, there are only five adjustments where the difficulty has gone down since July 2015.

Chapter 7:

Bitcoin Mining in Practical

a. **Current bitcoin mining situation**

I t is no secret that the Bitcoin cryptocurrency system together with its ubiquitous Blockchain technology ushered in an entirely new frontier, not just of financial freedom but of occasionally outrageous profits. Those who developed initial interests in Bitcoin mining namely cryptographers, cypherpunks, technically-minded and assorted hackers, were the first users stake their claim in a period that is called "gold rush." As of writing this, Bitcoin miners had collectively earned more than $2B from mining alone.

But is there any gold remaining that should be "mined?"

Here is what we know—Bitcoin mining has grown exponentially from a handful of early tech-enthusiasts into a full-blown industry, to a specialized industrial-level business.

The easy money that you could have scooped was taken a long time ago. What has remained buried under the complex mathematical cryptographic equivalent of the hard rocks. Transactions in bitcoin are usually stored in form of blocks by miners. When a miner successfully solves the complex mathematical hash problem attached to the block, he gains a payout as a reward. This payout is halved every four years. As of writing this, the current reward is 12.5 BTC per block.

But here is the twist to this: Bitcoin miners must compete with each other using computational power to validate the transaction on the Blockchain. This means that users with specialized, high-powered hardware can profitably extract bitcoins these days. But don't get me wrong here. Mining is still technically possible for any user, but those without massive processing power will find out that money is spent on electricity that will be generated through mining. Put simply, mining will not be profitable at a small scale unless you have access to cheap or free electricity.

Chapter 7: Bitcoin Mining in Practical

In the Bitcoin's early days, users mined the cryptocurrency using their home computers. Today, this isn't possible. Server farms that are made from thousands of custom-built hardware around the globe compete with one another validate the transactions on the Blockchain. Large companies that have the necessary resources to mount these servers have been allowing users to perform cloud mining and in return, users pay some fees for maintenance.

The transaction fees have historically formed part of miners' revenue. However, they've increased over time as the number of transactions approaches the bitcoin network's limit. As a result, users are willing to pay higher transaction fees to ensure that their transactions are fast-tracked on the Blockchain by miners. The debate about how to raise the limit has been at the heart of "deep divisions" that has divided the bitcoin universe.

b. What is hashing?

i. How to calculate your hardware hashing rate

A hash function is essentially a computer program that creates a summary of the large amounts of information that needs to be stored; this enables the system to store copious amounts of data without overwhelming its memory. A cryptographic hash function, which is what is utilized in blockchain, is a unique type of hash function that that converts an input message into a string of alphanumeric values. The value produced is known as the hash value or the digital fingerprint. The probability of two separate input messages creating the same hash value is extremely unlikely.

c. Desktop or graphics card based mining

If you're using your general purpose PC to mine today, you'll be disappointed with mining process. This is because, CPU mining—which allowed earliest miners to mine from the comfort of their homes—is no longer profitable with the current mining difficulty. Desktop mining (GPU or graphics card mining) is the second generation mining that began when users started to get frustrated with their CPUs. Instead of the CPU, GPU mining uses the graphics card, or the graphics processing unit.

Almost all modern PCs have an inbuilt GPU on their motherboards that support high-performance graphics. These cards are designed to improve the throughput in the PC and increase parallelism, both that is useful for Bitcoin mining. Bitcoin mining can easily be parallelized because you're trying to compute multiple hashes simultaneously with different nonces.

A language called the OpenCL—general purpose language for performing things other than graphics— was conceived in 2010 to help run many computational types faster on GPUs. This language paved the way for GPUs to be used for Bitcoin mining. At the time, mining with GPUs offered a number of attractive properties. First, the GPUs were easily available and therefore easy for beginners to set up. You could order graphics cards online or even purchase them at most of the big consumer electronics stores. Second, they allowed parallelism to be implemented for simultaneous SHA-256 computations. Some GPUs have specific instructions that allow them to perform bitwise operations which are essential for the SHA-256 algorithm. Others can be overclocked, allowing you to run transactions faster than they have been designed to function. The overclocking property is what was relied upon by gamers. With Bitcoin mining, GPUs provided a profitable mechanism for earning money.

Finally, you can drive many GPUs from one motherboard and the CPU. Therefore, you can take your PC that runs your actual Bitcoin node to gathers transactions from the Bitcoin network and assembles the blocks. As you assemble the blocks, you can attach multiple GPUs to it to attempt to get the right nonces that make the SHA-256 of the block correct.

Despite the advantages outlined above, GPU mining had some demerits such as:

- GPUs consume a fairly large amount of power, therefore a lot of electricity is used relative to the PC.

- Majority of the GPUs requires one to build their own boards or buy expensive boards that can house multiple cards which is a tedious process.

- GPUs have a lot of hardware that is built into them for performing video processing—floating point units—which can't be utilized for mining.

d. ASIC mining

On a really high-end GPU with aggressive tuning, you can get as high as 200 MH/s (200 million hashes per second). This hash rate is better than you can get with a normal CPU. But even with the improved performance, and even if you're an enterprising entity with multiple GPUs together, it would still take a longer time (300 years to be precise!) on average to find a block using the difficulty level in January 2015. Therefore, it's no longer in doubt that GPU mining is dead for Bitcoin mining today, even though it still crops up in some Altcoins.

Today, bitcoin mining is dominated by ASICs (Application Specific Integrated Circuits). ASICs are chips that have been conceived, built, and optimized for the primary purpose of mining the bitcoins. There are a couple of big vendors that sell these circuits to clients with a great deal of variety. You can choose between the slightly larger and more expensive models, to more compact models, as well as other models that have varying performance and power consumption levels.

Designing the ASICs requires a great deal of expertise and their lead-time is also quite long. Nonetheless, the ASICs can enhance the throughput in bitcoin mining. Up until December 2014, the lifetime of the ASICs was short due to the rapid increase in the bitcoin network hash rate, with the majority of the boards in the early ASIC days growing obsolete in about 6 months. During this time, the bulk of the bitcoin profits are made up-front.

Often, miners would make half of the expected bitcoin profits for the lifetime of the ASIC during just the first 2 months. This meant that shipping speed also became a crucial factor in breaking even. As a result of the immaturity of the bitcoin industry though, clients often experienced shipping delays with the boards becoming nearly obsolete by the time they arrived.

As the expansion rate of the Bitcoin's hash power stabilized, this mining equipment elongated their lifetime. At present, ASIC mining is the only practical means to be profitable in Bitcoin mining. However, this approach isn't friendly to small-scale miners because of the high costs involved.

Benefits of using ASIC mining hardware over conventional desktop-based rig

As you can recall, mining began on CPUs and then it shifted over to GPUs, and finally to ASICs. The ASICs are perhaps the last stop as far as the hardware is concerned for bitcoin mining. Here are some pros of the ASIC mining:

- They have faster hashing rates than GPU

- They can easily be scaled up for more hashing

e. Solo vs. Pool mining

One of the first questions that any user interested in mining bitcoin faces is whether to mine solo or join a pool. In solo mining, you decide to mine by yourself. In pool mining, you join a team and contribute your computing power to enhance your chances of getting rewards. There are multifold reasons both for and against each approach of bitcoin mining. But, if the hashpower distribution across the bitcoin ecosystem is anything to go by, the majority of bitcoin miners are opting for pool mining as opposed to solo mining.

If you're deciding on whether to join a mining pool or not, it is important to think of such a venture like a lottery syndicate. In a lottery syndicate, the merits and demerits are exactly the same. Going the solo way means that you won't have to share the bitcoin reward. However, the odds of getting that reward are significantly reduced. Even though a pooled mining has a much larger likelihood of solving the block and winning the bitcoin reward, that reward will always be split by all the pool members.

Thus, joining a pool generates a steady stream of bitcoin income, even if each bitcoin payment is meek compared to the full reward. It is also vital to note that a mining pool shouldn't exceed over 51% of the total hashing power of the bitcoin network. If a single group or entity ends up controlling more than 50% of a bitcoin network's hash power, it could – in theory – wreak havoc on the whole bitcoin system.

In January 2014, many bitcoin miners voiced concerns that GHash.io—bitcoin mining pool— was approaching the threshold, and the miners were advised to leave the pool. When choosing which mining pool to join, you should weigh up how each mining pool shares out its payouts and what fees (if there are any) that it deducts.

There are several schemes by which mining pools can divide the payments. Most of these schemes concentrate on the volume of shares or stake that the miner has submitted as a method to distribute the payouts. The most basic version of splitting the payments the typical "pay per share" (PPS) model. Variations on this model place limits on the rate that is paid per share; for instance, the Equalized Shared Maximum Pay per Share (ESMPPS) model, or the Shared Maximum Pay per Share (SMPPS) can be used. Pools can also prioritize payouts for how recently the miners have submitted shares. For instance, the Recent Shared Maximum Pay per Share (RSMPPS) can be used.

Chapter 8:

Real World Use

As a currency that has been around for less than eight years since it was first brought into existence, the Bitcoin technology can still be considered as one that is in its early stages. Even though there is already a highly-engaged society of early Bitcoin adopters, most of whom have performed a profound scientific analysis of the security risks of the Bitcoin protocol, there might still be several inherent defects or imperfections in the Bitcoin design. This may lead consumers to discard Bitcoin as a currency in favor of other currency designs.

Bitcoin has laid the path for people to accept digital currencies as a feasible alternative for numerous types of transactions in the Internet world.

This chapter will discuss just some of the products and services that you can pay for using your Bitcoins. Each place has its own terms for what charges are used when you pay for items with Bitcoins; fortunately, those charges are typically lower than what you would spend on traditional transactions or with other procedures.

a. Investing with bitcoin

Another thing that you can do with Bitcoin is using it as an investment. What this means is that you don't touch the Bitcoin but rather allow its value to increase, not unlike how you would enable investments in your 401(k) or other investment portfolios to increase in value.

While CryptoCurrency is new and there is certainly no guaranteed way of figuring out what the value of currencies such as Bitcoin will be in the long term, Bitcoin's value has increased from one ten-thousandth of a penny in the summer of 2009 to $3300 in the summer of 2017. If this trend continues, Bitcoin will be worth hundreds of thousands of dollars in a few years. Investments that have shown a lot of gain in a short period are usually seen as risky. However, in combination with other investments, Bitcoin can be a meaningful way to build value to your portfolio.

b. Shopping with bitcoin

Chapter 8: Real World Use

One of the largest retailers that accept Bitcoin payments is Overstock, which has a wide of products for sale including home furniture, electronic gadgets, and even jewelry. The prices you will see on the website are in US dollar, but you will see the option to pay using Bitcoins (BTC) on their checkout page. They offer global delivery to more than one hundred countries around the world. But as of this writing, only US citizens and residents can purchase from Overstock using Bitcoins. This could change once more places begin to see just how viable and exciting the Bitcoin may be.

Shopify is a well-accepted online commerce platform that started giving its online sellers the option to accept Bitcoin payments before the end of 2013. Shopify is composed of more than 70,000 online shops, but currently, there are only around 75 online stores that accept Bitcoin payments. You can visit Shopify's blog to see the list of those vendors. But the current number of 75 can potentially increase to more than 1,000 which is only a little over 1% of the 70,000-seller base of Shopify.

Other online stores that accept Bitcoins include Tiger Direct, Bitcoin Shop, Bitcoin Store and Memory Dealers.

You can visit Bitcoin blogs and websites such as Coin Map, Spend Bitcoins and Use Bitcoins to view updated lists of all online retailers who accept Bitcoin payments.

c. Withdrawal of bitcoin into cash

Traditional financial institutions can take days or sometimes even weeks to process a payment. The amount of time to process a payment is usually at least one business day. This means that the funds may appear in someone's bank account as if they are there, when in fact they have already been spent. Further, deposits to a bank account can take days to clear, meaning that the owner of those funds is not able to access them. This can be problematic when there is a need to make a critical purchase, such as healthcare. Instead of days, Bitcoin transactions take about ten minutes to process. Faster payment processing allows companies to run more efficiently because their funds aren't tied up in limbo. It also decreases frustration and increases peace of mind for customers.

Chapter 9 :

Blockchain In-depth

a. Introduction

Ablockchain is a database that holds a series of records in a list that is ever expanding. Each record is called a block and the blockchain is fully secured from any form of tampering and from revision. Each of the records, or blocks, is time stamped and is also linked to the record before it, hence the chain.

We all know of blockchains as being the underlying technology of the virtual currency, the bitcoin. The idea of the bitcoin was born in 2008 and in 2009 it became a reality and the blockchain is actually a public ledger of every bitcoin transaction. In the case of the bitcoin, each client can connect to the network and can send transactions to the network. They can also verify transactions, and take part in the heavy competition to make new blocks, a competition known as mining.

b. Structure of a block

Blockchains, as can be logically understood, consist of blocks that are verified and authenticated computationally before being added onto the chain. Each block has the following parts:

- A list of transactions

- A block header

Each block header, in turn, has at least the following 3 sets of metadata:

1. Information regarding the transactions that are in the block

2. The data and the timestamp regarding proof-of-work algorithm

3. A "hash" reference to the parent block (or the preceding block)

c. Block header

Whenever you make a bitcoin transaction, it is not appended to the Blockchain immediately. Rather, it is held in a memory pool that is also called the transaction pool. Some of the information in the transaction pool is held in a block header. Each block begins with a block header. The major version defines the block header parsing rules (which in this case is the block header format) and is incremented with each block header format update. Each block header has the following fields:

- **Version**. It describes the data structure inside the block. It is used so that nodes can read the contents of each block precisely and in the correct format.

- **Last Block**. It is an identification number of the previous block. It is important if you want to establish one of the current candidate blocks.

- **Merkle Root**. All of the bitcoin transactions inside the block are usually hashed together to form a single line of data. All of the fields must be unique, however, it would be fair to imagine a block as the most significant component of the block header. That is why the Merkle root is key—to help the fields remain unique.

- **Time**. It displays the current time.

- • ***Target***. It is a value that bitcoin miners use to add candidate blocks to the Blockchain. It is usually set by the bitcoin system.

To add any candidate block to the Blockchain, the data in the block header must be hashed and matched with the result of a certain target value. The target value is computed from the mining difficulty, which is a value that has already been set by the bitcoin network to regulate how difficult it is to append a block of transactions on the Blockchain. You can think of the target as some sort of "limbo" pole for the candidate blocks. In other words the greater the mining difficulty, the lower the target value, and the more difficult it will be to find the block hash that is below the target value.

d. Block identifiers

The main identifier of any given block is its hash—which is a digital fingerprint—obtained by hashing the block header twice using the SHA-256 algorithm. The resultant 32-byte hash is known as the block hash but strictly speaking, it's just the block header with the block header hash and only the block header can be used to compute

it. For instance,
000000000019d6689c085ae165831e934ff763ae
46a2a6c172b3f1b60a8ce26f is usually the block
hash of the initial bitcoin block that was ever
generated. The block hash helps to identify a
block uniquely and in an unambiguous manner
that can be independently derived by hashing the
block header.

It is worth noting that the block hash isn't
included in the block's data structure, neither is
the block transmitted on the bitcoin network.
Also, the block hash isn't stored on the node's
storage as a component of the blockchain. Rather,
the block hash is calculated by each computer as
the block is generated from the network. It might
be stored as an independent database table that
forms part of the block's metadata. This
information is necessary to fast-track retrieval of
blocks from the disk.

The second method to identify any block is to use
its position in the blockchain. This is usually the
block height. The first block that was ever
generated is at block height zero (0). This block is
the same block that was hitherto referenced by
the using the following block hash:
000000000019d6689c085ae165831e934ff763ae

46a2a6c172b3f1b60a8ce26f. A block can, therefore, be identified using two mechanisms:

- By referencing its block hash

- By referencing its block height

Each subsequent block that is appended on top of the first block is always one position ahead or higher in the blockchain. You can think of this as boxes that are stacked up one on top of the other. For instance, if the block height as at January 1, 2014, was roughly 278000, meaning. This means that there were 278000 blocks that were stacked up on top of the first block—which was generated in January 2009 (the first block).

Unlike the block hash, the block height isn't a unique identifier. Even though a single block has a specific and invariant block height, the reverse isn't necessarily true. The block height doesn't always identify a single block in the chain. 2 or more blocks can have the same block height that is competing for a similar position in the blockchain. Also, the block height isn't part of the block's data structure because is not stored inside the block.

Each node dynamically identifies the block's position (which in this case is the height) in the blockchain when it is generated and received from the bitcoin system. In this regard, the block height can be stored as metadata in an indexed database for faster retrieval.

e. SPV: How does Simple Verification works

The first SPV was conceived by Satoshi in his bitcoin whitepaper as a high-level design for implementing the bitcoin network. However, it wasn't implemented until 2011 when Mike Hearn developed BitcoinJ. The earliest SPV implementations were naive in the sense that they simply downloaded the entire blockchain, an approach that wasn't efficient compared with the full node if you use the bandwidth as a parameter.

By dropping bitcoin transactions that aren't relevant to the SPV node's wallet, the system was able to gain a significant disk usage savings. When BIP 37 was published in 2011, it offered a specification for Bloom filtering of the bitcoin transactions, which relied on the block header's Merkle root to authenticate inclusion of transactions just as had been proposed by Satoshi

Nakamoto in 2009. The implementation of bloom filtering in SPV greatly enhanced the bandwidth utilization in SPVs.

Ideally, when an SPV client syncs its data with the network, it links up to one or more validating bitcoin computers, finds out the latest block (at the tip of the blockchain), then requests all the block headers with the "getheaders" command to facilitate synchronization of blocks from the last block in the chain. If the SPV node is only interested in certain transactions that correspond to the wallet, it will generate a Bloom filter that is based on all the addresses for which the wallet owns the private keys and send the "filterload" command to the full client(s). This helps to return only the transactions that match the filter.

Now, after syncing the block headers—and possibly loading the Bloom filter—the SPV nodes transmits a "getdata" command to request every block that they missed out during synchronization for the last time. Once the node is in sync and remains linked up to the full node it will only receive the inventory messages for bitcoin transactions that correspond to the loaded bloom filter.

From a client's perspective, bloom filtering presents an efficient approach to determine relevant transactions in the chain, while utilizing minimum bandwidth, CPU resources, and disk space. Every block header in the blockchain will be 80 bytes. As of July 2017, there were only 38MB of data for the entire 8-plus year history of the entire blockchain. This means that each year, approximately 52,560 blocks—which corresponds to about 4.2MB—of data appended to the blockchain.

f. Benefits of Merkle root than simply hashing

As you've read in the previous section, each block on the blockchain contains a summarized list of all the transactions. Once the block is appended to the blockchain it becomes an immutable record (permanent record). This also means that if one bitcoin transaction is already present in one block, then there is no way it will appear in any other block of the blockchain. What makes this to be possible is a Merkle tree.

I know you're now asking, "What is a Merkle tree?"

Well, a Merkle tree—also called the hash tree—is a tree where every leaf node is labeled with block data and every non-leaf node has a hash value of the labels of the child nodes. Merkle trees are used in bitcoin transactions to allow secure and efficient verification of large transactions in bitcoin network. The root of the tree becomes the topmost computer and therefore the tree will be represented upside down while the bottommost nodes form the leaf nodes. Each node within the Merkle tree is simply a hash value of a transaction

Merkle trees are extensively used by the SPV nodes. The SPV clients don't have all bitcoin transactions and don't download the full blocks, but just the block headers. To verify that a transaction has been included in any block without downloading all the bitcoin transactions in the block, they use the Merkle path—which if the authentication path.

Let me use an illustration.

Suppose for instance, that an SPV node is interested in the incoming payments to an address that is contained in its wallet. Now, the SPV client will generate a bloom filter on its connections to nodes to restrict the transactions obtained to only those containing the addresses of interest. When a node sees a transaction that corresponds to the bloom filter, it will transmit that block using the "merkleblock" message. The merkleblock message has the Merkle path and the block header that connects the transaction of interest (incoming payments) to the Merkle root inside the block.

The SPV node can now use the Merkle path to link up the transaction to the block and authenticate that the transaction has been included in the block. The SPV client also uses the block header to connect the block to the rest of transactions in the blockchain. The combination of these 2 links, between the block and the transaction and between the blockchain and blocks, authenticates that the bitcoin transaction has actually been recorded in the blockchain.

Ultimately, the SPV client will have received less than a KB of data for all the transactions in the block header and Merkle path, an amount of data which is more than a thousand times less compared to the full block that would have been downloaded.

Chapter 10:

Cryptocurrency as the New Generation of Currency

a. Credit card and it's insecurity

It worth to mention that are many different types of Bitcoin ATM-s, and some are only operate in one way, however, some has a two-way function. It also depends where you use the Bitcoin ATM-s , as if you want to convert your existing Bitcoin to cash, you will probably receive the local currency of that country. Other issues that you might encounter is the fees. There are Bitcoin ATM-s that are operating with no fees, however some others can take as much as 5-10% fees once used. The legacy of Bitcoin is that are no fees, however, there are many operators who paid for producing such machines, as well costs to pay the electricity bills, rental fees, therefore some might charge for a certain fee. It is advisable to check the fees before using one, however the comfort that it provides are extremely helpful.

BITCOIN: FROM BINARY TO GOLD

How to use Bitcoin ATM-s

In case you wonder how it works, I can tell you from experience that is relatively easy. Let's assume that you want to buy some Bitcoin, using traditional Cash such as dollar. The checklist that you should have is this:

- Smart phone with internet connectivity: Any types of smart phones are ok.

- Hot wallet downloaded on the smart phone: Blockchain wallet or Jaxx

- Dollar bill: Ten, twenty or any dollar bill that you want to convert into Bitcoin.

- Bitcoin ATM near you: You can find a local Bitcoin ATM near you by visiting this link: https://coinatmradar.com/

Once you have downloaded one of the wallets I have recommended, or any other Cryptocurrency wallet to your smart phone, visit a local Bitcoin ATM.

In case you think that Bitcoin ATM-s are placed in some dark hidden street, let me tell you that normally the Bitcoin ATM-s are in a public place such as restaurants, pubs, or local shops, places that are many people visit daily.

b. Bitcoin transaction and it's security design framework

It's undoubtedly safe to assume that the bitcoin cryptocurrency is here to stay. Even though it is volatile, just like other cryptocurrencies, it has been around for much longer—a testimony of its resilience. But this doesn't mean that you should jump on the bitcoin bandwagon without understanding its security design frameworks.

Aside from its high price of entry, a horde of events that have occurred the over the years since 2009 clearly shows that while the bitcoin protocol is secure, the wallets and other services used to exchange and store bitcoin may not.

Here's a string of security design frameworks about bitcoin that you should be aware of:

Encryption and the Blockchain

Just like other cryptocurrencies, bitcoin implements cryptography as a central component of the protocol to establish anonymously (pseudonymous) and decentralized currencies. It uses SHA-256 encryption for both its Proof-of-Work (PoW) consensus protocol and transaction verifications. However, the security of the bitcoin protocol lies in the blockchain—one of its fundamental properties.

The blockchain is essentially a chain of multiple ledgers or blocks that have the transaction history. The blockchain begins with the initial block (also called the genesis block) and transactions and the solved hashes append new blocks after the genesis block, generating a blockchain. At the core of bitcoin protocol is the consensus algorithm that uses PoW to authenticate transactions that are appended to the blockchain.

Double Spending

It's possible—in spite of the belief to the contrary—to hoodwink the blockchain and spend the same money (bitcoins) twice, an action that is often called double spending. There are a couple of ways that this can be achieved. For instance, if a merchant doesn't wait for the transaction confirmation, the bitcoins can be double-spent by malicious attackers by quickly sending 2 conflicting transactions into the bitcoin network.

Another method is to pre-mine one transaction into the block and then spend the same bitcoins, before releasing the block to the blockchain. However, the amount of computing power that will be required to succeed at this makes it less productive than just mining the bitcoins legitimately. However, with the coming up of quantum computers this could be a reality.

Bitcoin wallets

Bitcoins are stored in the wallets. However, unlike conventional systems such as PayPal accounts, the wallets don't actually keep the bitcoins themselves. Despite the number of different formats and implementations, generally, wallets contain the public key (used to receive bitcoins), just like a bank account number. They also contain a private key which is used to prove that you are indeed the wallet owner if you attempt to spend the bitcoins.

Because of the above underlying frameworks, the bitcoin protocol isn't secure enough, but this doesn't extend to all the websites and services that you'll be dealing with on the bitcoin network. In the next section, we examine ways to be "street-smart" with bitcoin:

#1: Dealing with common scam tactics

Below are common scam tactics that you'll likely to come across while using Bitcoins:

- ***Fake Bitcoin exchanges***. You can find a link on social media claiming to provide Bitcoin exchange services. This can be a marketing trick. To avoid such tricks, always verify the URL of the website. Make sure its address starts with "HTTPS." If it starts with "HTTP", then you can deduct that it is not secured.

- ***Fake Bitcoin Wallets***. The fake wallets are just scams for malware that are disguised to infect your computer to steal private keys and passwords. Just like the fake Bitcoin exchange websites, you should verify the authenticity of theirs. Does it start with "HTTPS" or "HTTP"?

Chapter 10: Cryptocurrency as the New Generation of Currency

- **Phishing Scams**. Phishing occurs when someone tries to trick you into believing that they are a trusted organization or website by providing you with a fake site. Typically, phishers can contact you through email or fake web advertisement. When you visit that site by mistake you'll get either a malware or even lose your Bitcoins through fake transactions. The best practice is: don't click any hyperlink or email link. Always go directly to the website address.

- **Ponzi scams**. These are fake websites that promise to double your bitcoins in an overnight. Or, they can be any similar outlandish claims. The fact is; these sites are difficult to spot. The best practice to spot these scams is to look out for a text like this—a referral link: domain.com/ponzi/?ref=12345

- **Cloud mining scams**. These scams can be tricky to spot because not all the cloud mining operations are scams. Some websites are completely legitimate. However many are scams. Just like the fake Bitcoin exchange websites, you should verify the authenticity of theirs. Does it start with "HTTPS" or "HTTP"?

#2: Escrow Services

Bitcoin transactions are transparent, secure and immutable. Once a transaction has been verified and appended to the Blockchain, it can't be erased. So, what will happen when goods or services have been purchased and the seller doesn't want to transfer the promised goods or services? The transaction can't be reversed.

Such problems can be resolved by Escrow services. These are financial arrangement systems where a third party company holds and regulates Bitcoin funds required for the two parties—the seller and buyer—involved in Bitcoin transaction. The escrow services help to make transactions more secure by keeping the Bitcoin payment in a secure escrow account that is only released when all the transaction agreements have been met and overseen by the escrow services.

Escrow services are important when dealing with large amounts of Bitcoins where a certain number of obligations have to be met before payment is made. These services eliminate all the legal jargon and enable secure Bitcoin transactions. The only problem with escrow services is that such services they assume the roles of banks making Bitcoins to be similar to conventional currency systems again.

#3: Multi-signature Transactions

For corporate transactions,--or transactions that demand a high level of security— you can use multi-signature addresses that use more than one key for encryption and decryption of data. Multisignature transactions refer to situations requiring more than one key to authorize any Bitcoin transaction.

Suppose you're working for a company that wants to accept Bitcoins as a form of payment. Now, for security reasons, your company doesn't want only one employee to have access to the Bitcoin wallet's password details. Any Bitcoin transaction must meet the approvals of multiple employees in the organization. Such a system will require a multi-signature transaction.

A multi-signature address is simply an address that is linked to more than one private key. The simplest multi-signature address is an m-of-n address that's associated with n private keys. Sending Bitcoins from the address must have signatures from at least m keys.

c. Traditional financing versus blockchain

Currencies, crypto or otherwise need to follow three basic rules:

1. They need to be difficult to produce (cash) or find (gold or other precious metals)

2. They need have a limited supply

3. They need to be recognized by other humans as having value

Using only Bitcoin (BTC) as an example, it ticks the boxes of all three of these characteristics:

1. Bitcoin uses complex computer algorithms in its production which take a lot of computational power, so it cannot be replicated easily or at a discount

2. There are a finite supply of Bitcoins - 21
    Million to be exact[2]. As of 2015, roughly
    2/3 of this number had been mined

3. There are hundreds of Bitcoin exchanges
    and Bitcoin is accepted everywhere from
    Subway to OKCupid

Where cryptocurrencies differ from traditional
currencies (also known as Fiat currencies) is that
they are not tied to any one country, nation or
institution (in most cases). There are no USA
Bitcoins, no Japanese Litecoins or anything like
that. They are decentralized.

Bitcoin was designed as a "deflationary currency"
- meaning over time its value will, in theory,
inherently increase. Unlike fiat currencies which
are inflationary and whose value will eventually
decrease. After all, in 1917, $1 was worth the
equivalent of $20.17 today. So the US Dollar is
worth 20 times LESS than 100 years ago. In other
words, if you continue to hold $1 over the course
of 100 years, you will be able to buy progressively
fewer and fewer items in exchange for it, whereas
with Bitcoin, the opposite will happen.

BITCOIN: FROM BINARY TO GOLD

As another real world example. On 22 May 2010, Laszlo Hanyecz made the first real-world cryptocurrency transaction by buying two pizzas in Jacksonville, Florida for 10,000 BTC. Today 10,000BTC is worth over $40 million.

Bitcoin was designed this way so that no single person (or government) could increase the supply of money, lowering the value of the money already in the market.

We also have to remember that fiat currencies that we know and love were not always the main players in the currency world. For centuries, Gold and other precious metals were seen as the most desirable currencies for day to day usage. It was not until governments could standardize and verify the metallic content of coins (and later paper bills) that they became the go to choice for citizens.

Legendary Economist John Maynard Keynes had this to say about inflation and inflationary currencies.

While Bitcoin has an air of uncertainty about it based on the decentralization principle - where the real potential lies is in seeing it from the opposite perspective. With no single body being responsible for the supply of money, it forces all players (government, businesses and consumers) to be transparent about their processes, lowering the risk of fraud or tampering. The transparency is ensured by rewarding miners for their efforts (in the form of coin). This single dominating factor is why so many investors are confident about the long term viability of the currency.

One common argument made by Bitcoin detractors is that as there is no government backing the currency, it could totally collapse in theory. However, we have seen these happen numerous times with fiat currency under scenarios of hyperinflation where governments can no longer ensure the value of their money and as such have to create an entirely new currency. Common examples include the German Weimar Republic in the 1920s, where the currency lost so much of its value, banknotes were used as wallpaper. Currently, the Venezuelan economy is on track to experience over 1000% inflation for the year, leaving many citizens unable to afford daily necessities like bread. Bitcoin enthusiasts see the cryptocurrency as recession-proof.

The cost of international transactions is another area where cryptocurrencies maintain a huge advantage over traditional ones. Anyone who has ever had to send money overseas will know that the cost of processing these transaction can reach ridiculous levels. There are times when these fees can top 10%. As cryptocurrencies do not view international transactions (as there are no "nations" involved) any differently from local ones, there are minimal fees for sending money to any part of the world. The speed of transactions across borders is also much faster than regular fiat currencies, a Bitcoin transaction takes around 10 minutes to register as opposed to days for international bank transfers, and other coins process transactions even faster.

d. Government legality

Generally, cryptocurrency is legal. After all, it is not regulated by any central government. Hence, it cannot be restricted or controlled by nature. However, this does not mean that a government cannot restrict its use. For example, in countries like Ecuador and Bolivia, the use of cryptocurrencies is considered illegal.

It is also not hard to understand why governments are regulating the use of cryptocurrency. If you look at the features of cryptocurrency, especially its character of being anonymous, you will easily realize that it is a key to committing criminal activities like money laundering and others. Also, the use of cryptocurrencies can have an adverse effect on the value of legal tender currencies of the government.

The good news is that many states these days are being more acceptable of cryptocurrencies. As can be expected, governments tend to regulate its use. However, cryptocurrencies remain the same, and the blockchain technology remains unaltered. No government can have any influence its algorithm or be able to exercise control or influence over the network.

As of September 2017, the cryptocurrency industry is continuously being developed, and more individuals, businesses, and governments are being open to it. In fact, cryptocurrencies, especially Bitcoin, are now famous worldwide.

e. Government framework to regulate

There is currently no regulation of Bitcoin from the United States government or any government. This is because Bitcoin is a fairly new concept and it is designed to work more like something that people own instead of something that people use as a way to pay for things.

There is always a chance that the government will begin to regulate Bitcoin but it generally takes over a decade to regulate different forms of currency. In the past, it took a long time to make the switch and then the government was not able to regulate two different types of currency so they had to choose one. The chances of Bitcoin being chosen are very slim.

When the government makes the decision to regulate Bitcoin, they will need to figure out how they are going to make it less anonymous and they will have to make sure that they are prepared for the pushback that will come from the people who currently own Bitcoin.

It is a product that can be traded, not something that can be spent.

Chapter 11:

The Future of Bitcoin and Other Cryptocurrency

The Blockchain technology exposed the world to the concept of trustless data structures giving us a sneak-peek of the future with the launch of Bitcoin cryptocurrency. The Bitcoin cryptocurrency system has indeed changed the currency landscape as we've known it. In this chapter, we explore the future of the Bitcoin and other cryptocurrencies. Let's dive in.

a. Global acceptance and mass adoption

For some time, Bitcoin and other cryptocurrencies have been the domain of 4 broad categories of crypto users: software developers, criminals, speculative investors, and libertarians. The average person has been dissuaded from adopting bitcoins and other cryptos due to a number of factors such as:

- Lack of technological understanding of bitcoins and the whole concept of cryptocurrencies.

- The challenging prospects of downloading wallets with many reports of hacking the exchanges and price volatility.

- Emergence of Ponzi schemes, especially in cloud mining

- Generalizations that bitcoin and other cryptos are currencies for terrorists and money launderers operating on the dark web.

As such, the gap between the crypto-haves and crypto-have-nots (fiat currency users) is big. As of writing this book, less than 1% of the total global population owns cryptocurrencies, even though there are 3.8 billion people that have access to the internet and 5 billion people with bank accounts that rely on fiat currencies that could enormously benefit cryptocurrencies.

Some of the limitations that cryptocurrencies presently face – such as the fact that one's digital fortune can be erased by a computer crash, or that a virtual vault may be ransacked by a hacker –

may be overcome in time through technological advances. What will be harder to surmount is the basic paradox that bedevils cryptocurrencies – the more popular they become, the more regulation and government scrutiny they are likely to attract, which erodes the fundamental premise for their existence.

Before clients can conduct any significant volume of transactions through cryptocurrency markets, the challenge of liquidity—extent to which the market allows assets to be purchased and sold—must be overcome. For instance, Bitcoin can drop in prices to more than 10% in a day, a fact that doesn't instill confidence in the global market.

However, volatility is increasing in the main cryptos as their prices continue to shoot up and their markets grow. As of writing this, the Bitcoin was exchanging with US Dollar at the rate of 1BTC for $8200. The increase in the stability will be further enhanced when users pay for everyday transactions using these cryptocurrencies with the emergence of debit cards.

To improve the adoption rate among the masses much more should be done to make the cryptos to be user-friendly and more democratic. The good news though is that a wave of revolution is now taking place that will begin easing users into the crypto universe with greater assurance. A couple of blockchain startups have begun to implement systems that bridge the gap between the crypto universe and classical currencies to be possible. Therefore, going into the future, we expect the number of users joining the cryptocurrency bandwagon to increase.

b. Fixing Supply and growing demand

The bitcoin crypto celebrated and admonished in equal measures by supporters and skeptics, partly because of its finite supply. There are only 21M bitcoins that can and will ever be mined, irrespective of the world's population and the corresponding demand for them. Once all the 21M have been mined, there won't be more bitcoins (not unless the protocol is altered to increase the supply).

Bitcoin enthusiasts love the fixed supply concept it harkens back the crypto to the days of the sound money as a medium of exchange comparable to gold standard. As a matter of fact, gold shares many of its similarities with Bitcoin with the most obvious one being fixed supply. As you know, gold can't be generated out of the thin air in arbitrary amounts. It has to be extracted from the earth and placed into circulation as the market prices demand.

The gold standard property hinders banks and other organizations' abilities to offer fiduciary media because at some point these organizations can be forced to redeem their paper notes in gold. Bitcoin can accomplish these properties because of its own fixed supply. Bitcoin also takes the gold's merits a step further, even though, it is virtual. The Bitcoin supply isn't only unable to being arbitrarily manipulated, but it also eliminates the need for using paper substitutes, thus it is weightless and cost less to store.

Gold is so heavy and takes up so a lot of physical space, thus users under a gold standard will tend to use paper substitutes rather than carrying the gold. This leaves users with banks as the only institutions that can handle them responsibly. Essentially, they have to trust these institutions for proper and safe storage, something that Bitcoin has solved through decentralization.

Despite the promising benefits of Bitcoin, the user still has a problem with the Bitcoin's finite supply.

For instance, how will bitcoin miners fare once they lose the block rewards? Bitcoin miners worry that mining will become unsustainable since all the bitcoins will have been generated. This will only leave transaction fees as means for financial sustainability. Critics argue that relying on transaction fees rather than block rewards will be unaffordable, leading to a contraction of miners and centralization of the system leading to a collapse.

The result of the discrepancies between the supply of and demand for bitcoins will be a steady and gradual decline in the price level that parallels to an equally steady increase in the purchasing power of the currency. Thus, as the bitcoin miners collect transaction fees over time, the bitcoins will gain value. This value appreciation across the time will turn the fee-centric mining into financially infeasible tasks to sensible and long-term investments.

However, the Bitcoin's vision shouldn't be limited by this discourse. The inability to imagine of the future doesn't render it unfeasible; the impulsive evolution and shifting of the market economy in the age of fast-paced, complex IT should remind us of this basic fact every day.

c. Other type of cryptocurrencies

This is the decade of the cryptocurrency. Bitcoin's value has skyrocketed—to $8200, well past the price of 1 ounce of gold. Equally surprising is the fact that multiple new cryptocurrencies like Ethereum and Zcash have also jumped into the bandwagon, bringing the total market capitalization of all the cryptos to over $200B with Bitcoin controlling more than $100B as of writing this book.

But the long-term viewpoints on cryptos has also gotten blurrier. After years of dominance by Bitcoin, squabbles among the core developers and miners punctuated by lack of progress and rising transaction fees, Bitcoin's attractiveness as long-term investment and payment system has been eroded. Ethereum is a strong rival to bitcoin, but it's an entirely different crypto, with a focus on apps built on the blockchain (smart contracts) instead of the naive payments.

And new cryptos, promising new and enhanced features, seems to be launching every month. As of writing this, there are 1110 cryptos already registered on the Coinmarketcap.com with some having failed with others still trying to find their foot in the crypto universe. For sure, no one can tell with precision where to invest or what will happen to these cryptos in the future. However, this is a good moment to examine some of the cryptos with a bright future besides Bitcoin:

- Ethereum

- Litecoin

- Zcash

- Monero

- IOTA

- NEO

- Ethereum Classic

- EOS

- BitShares

- Ark

- Komodo

- SALT

- Bitcoin Gold

- Bitcoin Dark

- Bitcoin Cash

- Iconomi

d. Conclusion

The Bitcoin technology together with its ubiquitous Blockchain technology exposed the world to the concept of applying trustless data structures and in the process, giving us a glimpse of the future. Surely, the Bitcoin cryptocurrency has given us that better view if the wave of changes currently taking place in financial institutions is anything to go by. Iteratively, it presents a remarkable step forward, but we aren't there yet.

Even though the Blockchain together with the bitcoin crypto has become increasingly popular in the past few years, many people are still not sure exactly what it is. At present, less than 1% of the global population has access to cryptocurrencies, even though 3.8 Billion people have access to internet users. As such, the bitcoin crypto as envisioned by Satoshi Nakamoto as a P2P, trustless, decentralized currency with the primary aim of killing regulation is still work-in-progress.

Can the bitcoin crypto navigate this potential fork in the road and the myriad of challenges ahead? While the future is forever uncertain, the real question is whether, in Bitcoin's case, it will revolutionize the financial industry and adapt to the needs of the real world. My best bet is that Bitcoin will indeed navigate these challenges and become "the next-generation currency."

If you think that you have learn something out from this book or feel this book is worth the recommendation to the community do kindly drop a 5 star review on the product page.